DISPATCHES FROM PLANET 3

Also by Marcia Bartusiak

Thursday's Universe
Through a Universe Darkly
Einstein's Unfinished Symphony
Archives of the Universe
The Day We Found the Universe
Black Hole

Dispatches from Planet 3

Thirty-Two (Brief) Tales
on the Solar System, the
Milky Way, and Beyond

———————————

MARCIA BARTUSIAK

Yale

UNIVERSITY PRESS

New Haven and London

520
Bartusiak

26⁰⁰

Published with assistance from the foundation established in memory of Philip Hamilton McMillan of the Class of 1894, Yale College.

Illustration courtesy of Barbara Schoeberl, Animated Earth, LLC, for the cutaway of a massive star in Chapter 15.

Yale University Press books may be purchased in quantity for educational, business, or promotional use. For information, please e-mail sales.press@yale.edu (U.S. office) or sales@yaleup.co.uk (U.K. office).

Set in Janson type by IDS Infotech, Ltd.
Printed in the United States of America.

Library of Congress Control Number: 2018933935
ISBN 978-0-300-23574-6 (hardcover: alk. paper)

A catalogue record for this book is available from the British Library.

This paper meets the requirements of ANSI/NISO Z39.48-1992 (Permanence of Paper).

10 9 8 7 6 5 4 3 2 1

For Steve, my polestar

Contents

CONTENTS

CONTENTS

Preface

Almost no discovery in astronomy is a total surprise. Not really. Sifting through the histories of unexpected findings, one frequently encounters early hints, precursors, a sense of something in the air. As Isaac Newton famously noted in the seventeenth century, "If I have seen further, it is by standing upon the shoulders of giants." In other words, predecessors often pave the way, which makes the journey toward a new scientific vision possible.

My mission over the past several years, when writing the essays contained in this book, was to provide the back story for many recent astronomical discoveries. These explorations have included events in our nearby celestial neighborhood and out to the farthest reaches of the universe. Even beyond space-time to the multiverse.

I had no determined path in the course of my inquiries. Upon coming across a specific news item that piqued my interest, I headed to the archives to uncover a richer context. The controversial demotion of Pluto to dwarf planet, for example, reminded me when another solar-system member was similarly downgraded in the nineteenth century. And the discovery of an amino acid within a stream of interstellar dust is only the latest confirmation of our intimate connection to the cosmos, knowledge that is surprisingly recent. For most of history, astronomers were not at all sure that the stuff of the heavens was the same as the stuff on Earth. And when news stories kept referring to Edwin Hubble as the "discoverer of the expanding universe," I couldn't help but let readers know that a humble Belgian cleric, Georges Lemaître, and a former Midwest farmboy, Vesto Slipher, were equally responsible for revealing this astounding cosmic property.

My armchair investigations whisked me off in spirit to exotic locales: to ancient Mars, when liquid water once flowed freely on its surface; to an ensemble of galaxies that to our eyes resembles *Alice in Wonderland*'s Cheshire Cat; to the collision of two massive black holes a billion light-years away, an event that released fifty times more energy than all the stars in the universe were radiating at that moment; and finally down to the limit of the smallest quantum grain, where space and time allegedly come unglued and start to wink in and out of existence in a probabilistic froth.

While assembling these articles into book form, I was pleasantly reminded: I hadn't kept track of how many women I had portrayed over the years. I always let the news set my agenda and, lo and behold, there they were. Vera Rubin brings dark matter to the forefront of astronomical concerns; Jocelyn Bell keenly spots a bizarre new star; Henrietta Leavitt ingeniously devises a revolutionary cosmic yardstick; Jane

Luu co-discovers the first solar-system object beyond Neptune and Pluto; Beatrice Tinsley proves that galaxies evolve; Cecilia Payne tries to reveal the universe's major elemental ingredient (until told to ignore it); and Margaret Burbidge contributes the observational proof that the calcium in our bones, the iron in our blood, and the oxygen we breathe came from the ashes of ancient stars. Many of these names are not found in astronomy textbooks, so it was gratifying to bring them into the spotlight.

These are among the thirty-two stories in this collection, which I have loosely arranged, starting with our solar system and working outward in space-time to the Big Bang . . . and beyond. There is no need, however, to read them in this particular order. For the most part, each chapter stands alone, so you are welcome to wander, as I have, along your own desired path among the stars.

DISPATCHES FROM PLANET 3

CHAPTER ONE

Earth Is but a Speck

Our cosmic address keeps getting longer

W ALK into an open field on a clear, moonless night. Overhead, sparkling stars are sprinkled across the sky. All of them seem equidistant from you—and no one else—and you are lulled into imagining yourself at the center of the universe.

For nearly five hundred years, astronomers have struggled to break that illusion. Our petty standing in the cosmos is a scientific fact, if not a visceral experience. Earth zips at nearly 67,000 miles (108,000 kilometers) an hour around the Sun, which in turn completes one lap around the Milky Way every 220 million years, meaning that the last time we were in this neck of the galaxy, dinosaurs were getting ready to rule the planet. Still, as you look skyward in that pitch-black field, Earth seems to be at the heart of all creation.

We should blame Aristotle for initiating that perspective. So authoritative was his pronouncement of an Earth-centered universe in the fourth century BC that few challenged the idea for nearly two millennia. But over time, the urge to better explain the universe's behavior gave rise to new concepts of Earth's celestial position. In 1543, Nicolaus Copernicus reintroduced a hypothesis first posited by the ancient Greek Aristarchus of Samos some eighteen centuries earlier. His model boldly placed the Sun at the center of the universe, shoving the Earth into motion. The radiant Sun was at last in its proper perch, "as if resting on a kingly throne," wrote Copernicus.

Copernicus was not disturbed at all by a moving Earth, frightening as that might seem at first. More disturbing to him was the rotating sky in an Earth-centered universe. The farther out one moves from a stationary Earth, the faster and faster the sky must move to stay in place. But the Polish mathematician and astronomer knew quite well the consequences of challenging conventional notions. In the preface to his great work *De revolutionibus orbium coelestium* (On the revolutions of the heavenly spheres), he predicted that "as soon as certain people learn that in these books of mine . . . I attribute certain motions to the terrestrial globe, they will immediately shout to have me and my opinion hooted off the stage."

That misfortunate fate fell upon Galileo, who starting in 1609 gathered the crucial evidence supporting Copernicus's heliocentric vision. In 1633 he was brought before the Inquisition and eventually put under house arrest for daring to oppose an Earth relaxing at the universe's center.

By the time of Newton decades later, though, such hostility had finally faded. For one, Sir Isaac's physics could at last explain why we aren't thrown off the planet as the Earth rotates and orbits the Sun. Yet even though Copernicus moved Earth from the hub of the solar system, its inhabitants

Earth as seen from Apollo 11 in 1969.
(NASA)

remained confident that they retained a privileged place at the center of the Milky Way, then thought of as the sole galaxy. *Homo sapiens* is an egotistical species; we resist being kicked out of a prime spot in the cosmic scheme of things.

That confidence, though, withered as astronomy underwent a spectacular transformation starting in the nineteenth century, an era teeming with technological innovation. Prominent industrialists, enriched by the Gilded Age, provided the money that allowed dreamers to construct the powerful telescopes they had long desired.

With one of those new instruments atop California's Mount Wilson, Harlow Shapley resized the Milky Way. He discovered in 1918 that it was ten times larger than previously thought, and along the way, he relocated the Sun and its planets into the galaxy's suburbs. The Sun resides roughly 30,000 light-years from the galactic center, more than halfway to the Milky Way's edge. "The solar system is off center, and consequently, man is too," Shapley liked to say.

But Shapley did not take the next step; he, too, fell victim to cosmic pride. Despite the growing circumstantial evidence that the Milky Way was not alone in the universe, he held fast to his beloved "Big Galaxy" model. In this scheme our galaxy remained at center stage, meaning we lived in a solitary, star-filled oasis suspended in a darkness of unknown depth.

Shapley's vision was demolished in 1924, when Edwin Hubble at last proved that the cosmos is populated with myriad galaxies as far as the telescopic eye can see. The Milky Way suddenly became a bit player in a much larger drama.

As you can see, the history of astronomy is a continuing extension of the Copernican principle, moving us farther and farther from a front-row seat. It's a principle of irrelevance that involves not only our position in space and time but also the contents of the universe. In recent decades, astronomers have learned that a hidden ocean of cosmic matter—comprising about 85 percent of the universe's mass—surrounds us, possibly elementary particles yet to be discovered. The stuff of stars, planets, and us is but the flotsam in this enveloping sea.

More startling—and taking the Copernican principle of displacement to its ultimate end—our universe may not be the only cosmos. As physicists attempt to construct a theory that unifies all the forces of nature, one theme repeatedly arises: that additional cosmic realms may be lurking in other dimensions. We could be part of the multiverse; the Big Bang might have occurred when universes outside our dimensional borders bumped into one another.

The main response to this astounding theory has been to bury our heads in terra firma. Yet such a wider perspective can eventually offer soothing succor, allowing our earthly concerns to shed away slowly, until they dissipate completely. Hubble knew this. During a visit to the astronomer's home, the English poet Edith Sitwell was shown slides depicting the many galaxies that cannot be seen with the naked eye. "How terrifying!" she exclaimed. To which Hubble replied: "Only at first—when you are not used to them. Afterwards, they give one comfort. For then you know that there is nothing to worry about—nothing at all."

Granted, the hugeness of the cosmos is difficult to perceive and, as Sitwell expressed, horrifying to ponder. A character in Thomas Hardy's nineteenth-century novel *Two on a Tower* gives splendid voice to this apprehension: "There is a size at which dignity begins; further on there is a size at which grandeur begins; . . . further on, a size at which ghastliness begins. That size faintly approaches the size of the stellar universe," says astronomer Swithin St. Cleeve in the novel.

Indeed, our cosmic address is getting excruciatingly long: Planet No. 3, Solar System, Orion Spur on the Sagittarius

Spiral Arm, Milky Way, Local Cluster, Virgo Supercluster, Universe, Multiverse.

It's time for earthlings to acknowledge our minor-league status and collectively grasp the magnificent vastness that engulfs us all. While a widespread recognition of Earth's humble station is unlikely to end conflict here, fully comprehending our planet's infinitesimal place in the universe might be a modest step toward diminishing our hubris. Earth is but a speck, the cosmic equivalent of a subatomic particle hovering within an immensity spanning billions of light-years.

Yet, don't despair. We can still savor our cleverness in figuring out both this and the many other cosmic mysteries in the pages ahead.

Celestial Neighborhood

Well into the nineteenth century, astronomers spent much of their time devoted to our local celestial neighborhood. They aimed their telescopes at the solar system and prominent stars in the nighttime sky. The boundaries of the known universe then encompassed only one galaxy, our beloved Milky Way. And standing like a colossus over these astronomical endeavors, even after his death, was Sir Isaac Newton, whose momentous law of gravitation enabled astronomers to predict the motions of the Moon, planets, and comets. At the same time, their mathematical and observational diligence allowed them to detect new and unexpected objects within the solar system, such as asteroids. Such discoveries inevitably led to discussions over what is and is not a planet.

Better telescopes also led to sharper seeing. Early on, observers found a ring around Saturn and canal-like features on Mars. Could that mean there is water on the red planet? Is there other life in the solar system, or even on planets circling other stars?

And what about those stars? For most of astronomical history, the stars primarily served as a backdrop in studies of the solar system, but by the twentieth century that all changed as astronomers realized that stars come in a range of sizes, from huge red giant stars to tiny white dwarf stars no bigger than the Earth. And before astronomers could even get comfortable with that fact, they were faced with even weirder possibilities: the neutron star no bigger than a city and something even smaller and more bizarre.

CHAPTER TWO

Bedazzled by a Comet

How a comet validated the laws of Sir Isaac Newton

S OME repeatedly come and go with the precision of a
clock. Others arrive unexpectedly at our cosmic
doorstep, providing a few days or weeks of nighttime
entertainment, only to disappear into deep space and
never return. And a few fizzle out altogether.

For centuries, people both feared and revered comets. For
many they were harbingers of disaster, their long tails sweeping
across the sky like a fiery sword, symbol of death and destruction. But to others they were messengers of good news. Shortly after Julius Caesar's assassination in 44 BC, a comet appeared
in the sky that was so luminous it could be seen in broad daylight, a rare feat for a comet (it has only happened nine times in
the past three centuries). Caesar's successor, Augustus, wrote
that this brilliant star signified "that the Soul of Caesar was
received among the Divine powers of the immortal Gods."

Humanity had to await the Age of Enlightenment for a
more reasoned explanation of a comet's nature. It appeared in
the grand finale of the *Principia*, Isaac Newton's masterful

treatise on gravitation published in 1687. There, in his closing chapter, Newton laid out his mathematical theory of the motion of comets—an effort, he told a colleague, that was "the most difficult of the whole book."

Newton had been inspired by the appearance of a spectacular comet in 1680, the first comet to be discovered with a telescope. In the *Principia*, Newton traces the path of this comet across the constellations during the months it was visible. A diagram he included in his book was the first figure in astronomical history to show a comet completely swinging around the Sun, owing to gravity. Before that, observers were not sure that a comet approaching the Sun was actually the same object seen later to fly away from it. Newton had accurately determined that "comets are a kind of planet and revolve in their orbits with a continual motion." Their paths could be in the form of a very elongated ellipse, similar to a planet's, or an open hyperbola. In that case, the comet would forever depart from the solar system.

Newton also concluded that the comet was "solid, compact, fixed, and durable," just like the bodies of planets. "For if comets were nothing other than vapors or exhalations of the earth, the sun, and the planets," he wrote, "this one ought to have dissipated at once during its passage through the vicinity of the sun." And the tail? Hardly more mysterious than an "extremely thin vapor that the head or nucleus of the comet emits" when heated by the Sun. Comets were not omens of doom, Newton was saying. They were simply small planetoids. Nothing to be afraid of.

More problematic to his fellow scientists was Newton's law of gravity itself. His mathematics implied that imperceptible ribbons of attraction somehow radiated over distances, both long and short, to keep moon to planet and boulder to Earth. For many, this feat appeared more resonant with the occult than science.

The German astronomer Johannes Kepler in the early 1600s had suggested that threads of magnetic force emanating from the Sun were responsible for pushing the planets around. A little later the French philosopher René Descartes visualized the planets carried around like leaves trapped within a swirling whirlpool by vortices of *aether*, the tenuous substance then thought to permeate the heavens.

Newton's critics were now demanding a physical mechanism. What was replacing either magnetism or vortices? This led to Newton's famous statement in the *Principia:* "I have not as yet been able to deduce from phenomena the reason for these properties of gravity, and I do not feign hypotheses." Newton did not want to stoop to speculating or conjuring up some kind of hidden cosmic machinery. It was enough for him that his laws allowed successful calculations to be made.

Total acceptance took a while, but as the years passed, the rest of the physics community did eventually come over to Newton's side. And it was a comet, of all things, that provided the incentive.

Edmond Halley, Newton's colleague at the Royal Society of London, had used his friend's mathematical laws to make the first prediction of a comet's return. After poring over historic records, Halley had compiled a list of twenty-four comets observed from 1337 to 1698 and computed their motions. Looking over this record, he came to recognize that a comet sighted in 1682 had much in common with comets previously

A photo of Halley's comet taken at Yerkes
Observatory on May 5, 1910.
(Wikimedia Commons)

observed in 1531 and 1607. For one, they shared the same orbital characteristics (all went around the Sun in the opposite
direction to the planets). This made him suspect it was the
same comet returning every seventy-five to seventy-six years.
"The space between the Sun and the fixed stars is so immense," wrote Halley, "that there is room enough for a comet
to revolve, though the period of its revolution be vastly long."

Based on his calculations, which took into account the
additional tugs by Jupiter in the comet's journey through the

solar system, Halley made a prediction. "I dare venture to foretell," he announced in his 1705 paper, "that it will return again in the year 1758."

The comet appeared on schedule, just as Halley foretold. On Christmas Day in 1758, thirty-one years after Newton's death and sixteen years after Halley's, an amateur astronomer and gentleman farmer in Saxony named Johann Georg Palitzsch was the first to catch sight of the comet as a nebulous star in the nighttime sky. French observer Charles Messier, already on the lookout for the comet, saw the same fuzzy object four weeks later from Paris. It was soon confirmed to be Halley's returnee, and by March the comet was rounding the Sun.

The public was bedazzled, and the remaining critics of Newton's controversial law of gravity were instantly silenced. Despite the lack of a mechanism, his law was at last triumphant among both scientists and the public. Who could argue with a theory that allowed for a spot-on prediction about the solar system's behavior nearly a century in advance?

As a consequence, the universe came to be viewed as intrinsically knowable, ticking away like a well-oiled timepiece. And Halley's name became forever linked to that special, periodic celestial visitor. Its next visit: 2061.

CHAPTER THREE

To Be . . . or Not to Be a Planet

How planets have been promoted and demoted
over the decades

THEY looked for five years toward the far edge of the solar system and found nothing. But in 1992 the tide at last turned for two American astronomers. Using a new digital camera mounted on the University of Hawaii's 2.2-meter telescope atop Mauna Kea, David C. Jewitt and Jane X. Luu swiftly spotted their long-awaited quarry: a fuzzy spot of 23rd magnitude, four billion times fainter than the star Sirius. They had found the holy grail for planetary astronomers: an object orbiting the Sun beyond Neptune and Pluto. It was roughly 125 to 150 miles (200 to 240 kilometers) wide. They had wanted to name it "Smiley" (after the astronomer Charles Hugh Smiley), but given that an asteroid had already been named for Smiley, it's today simply referred to by its catalog name, 1992 QB1. That's astronomy code for being the twenty-seventh asteroidal object discovered in the second half of August in the year 1992.

Over the ensuing years, Jewitt, Luu, and others found many more objects like 1992 QB1 in the far reaches of the solar system, and they were christened with such captivating names as Quaoar, Sedna, Makemake, and Haumea. These newly discovered bodies were proof that the "Kuiper belt," a thick ring of icy planetesimals beyond the solar system's outer planets, indeed existed, as proposed in the mid-twentieth century by, among others, the Dutch-American astronomer Gerard P. Kuiper (although he originally thought the belt, remainders from our solar system's birth, would have scattered away by now).

During that time of explosive discovery, Luu made a prophetic remark in *Astronomy* magazine about the new evidence: "The confirmation of the Kuiper belt changes our perception of the solar system. What we thought of as a planet is probably just the biggest member of a rather large population of objects." She was thinking of tiny Pluto, only 1,400 miles (2,250 kilometers) wide. Smaller than our Moon, it had always been an oddball when compared with its gas-giant neighbors— Jupiter, Saturn, Uranus, and Neptune. The best ammunition to support this notion arrived when California Institute of Technology (Caltech) astronomer Michael E. Brown announced in 2005 that he and his colleagues had found an object in the belt heavier than Pluto. It has roughly a third more mass.

Brown's newfound body was eventually dubbed Eris, after the Greek goddess who personifies strife and discord. It was a fitting name, because this groundbreaking work caused the International Astronomical Union to revamp the solar system's membership. By 2006 Pluto was demoted in status to "dwarf planet," no longer in the big time but simply one of the larger members of the Kuiper belt, like Eris and the others. For their pioneering roles in this transformation of the solar system, Jewitt, Luu, and Brown were awarded the 2012

Kavli Prize in astrophysics, a prestigious biennial honor that came with a cash award of $1 million.

Many a child (and adult) was horrified when the number of planets in our solar system dropped from nine to eight. It means the Italian-menu formula for remembering their order is out: instead of "My Very Educated Mother Just Served Us Nine Pizzas," we have "My Very Educated Mother Just Served Us Nachos." A banquet of pepperoni and cheese has been reduced to an appetizer.

Since the eighteenth century, we've been accustomed to astronomers *adding* planets to our solar system, not subtracting them: first Uranus in 1781, followed by Neptune in 1846. It seemed an unprecedented move for astronomers to take one away: the planet found by Lowell Observatory astronomer Clyde W. Tombaugh and greeted with such fanfare in 1930. Pluto, we hardly knew ye. But this is not the first time the solar system has undergone a substantial reconfiguring. Another planet once came and went in a similar manner—two centuries ago.

Ever since Johannes Kepler, in the early 1600s, was able to link a planet's orbital period (the time it takes to round the Sun) to its orbital radius, astronomers sought an underlying pattern to the various distances of the planets from the Sun. In 1766 the Prussian scientist Johann Daniel Titius developed an elaborate mathematical scheme (based on earlier work by Oxford professor David Gregory) that appeared to account for the planets' positions. Six years later, a self-educated astronomer soon to be a professor at the Berlin Academy of Sciences, Johann Elert Bode, drew attention to the pattern in

The pattern from Bode's *On the New Eighth Major Planet
Discovered Between Mars and Jupiter*

Planet	Predicted mean distance from Sun	Observed mean distance from Sun
☿ Mercury	387 units	387
♀ Venus	387 + 293 = 680	723
⊕ Earth	387 + 2 × 293 = 973	1,000
♂ Mars	387 + 4 × 293 = 1,559	1,524
Probable planet between Mars and Jupiter	387 + 8 × 293 = 2,731	
♃ Jupiter	387 + 16 × 293 = 5,075	5,203
♄ Saturn	387 + 32 × 293 = 9,763	9,541
♅ Uranus	387 + 64 × 293 = 19,139	19,082

a new edition of a popular book on astronomy that he had written, which led to the rule becoming known as "Bode's law." The one shortcoming of the law was that it did not account for an apparent gap between Mars and Jupiter, where the law predicted an intermediate planet should appear.

When the planet Uranus was discovered at the very distance from the Sun that continued the sequence beyond Saturn, the sway of Bode's law (though not based on any physics) became near-mystical, immediately emphasizing the yawning gap between Mars and Jupiter. "Can one believe that the Creator of the Universe has left this position empty? Certainly not!" declared Bode. The success with Uranus encouraged astronomers throughout Europe to join forces to discover the planet everyone was sure was missing beyond Mars. The team jokingly referred to itself as the "celestial police," dividing the sky into twenty-four zones so each could be thoroughly explored by one of the team.

Meanwhile, the discovery of an object orbiting in the "gap" was serendipitously made by one of the astronomers the "police" had intended to enlist—although he didn't know it. Working from a new observatory he had founded in Palermo, Sicily, the monk Giuseppe Piazzi was assembling a star catalog, the most accurate in its day. On the evening of New Year's Day in 1801, he routinely measured the position of a star in the constellation Taurus, the Bull. "The light was a little faint, and of the color of Jupiter," he reported, "but similar to many others which generally are reckoned of the eighth magnitude. Therefore I had no doubt of its being any other than a fixed star."

But, following his customary procedure, Piazzi measured the star again the next night and found to his surprise that it had shifted. Over subsequent nights, he kept track of its movements and saw that its path was not elongated, like a comet's, but rather more circular. Privately, he wondered whether it might be the long-sought lost planet. "Since its movement is so slow and rather uniform," he wrote a colleague, "it has occurred to me several times that it might be something better than a comet. But I have been careful not to advance this supposition to the public."

By February, Piazzi was unable to continue his observations because the object was lost in the glare of the Sun, but he communicated his find to other astronomers. Although they could not observe the newfound body, the noted German mathematician Carl Friedrich Gauss was able to calculate its orbit from the limited data. That helped astronomers relocate Piazzi's object once it was again visible, on December 31, near the very spot in the constellation Virgo, the Virgin, that Gauss had computed. More than that, its orbital radius closely matched that predicted by Bode's law.

Piazzi named the object Cerere Ferdinandea (Italian for "Ceres of Ferdinand"), in honor of the patron goddess of

An image of Ceres taken by NASA's *Dawn* spacecraft in 2017.
(NASA/JPL-Caltech/UCLA/MPS/DLR/IDA)

Sicily and his own patron, King Ferdinand IV of Naples and Sicily. Bode excitedly wrote a paper in 1802 trumpeting the discovery (and not forgetting to crow about his own role in the endeavor): "Piazzi had, indeed, here discovered a very extraordinary object. It was most probably the eighth major planet of the solar system, which already thirty years before I had announced between Mars and Jupiter, but which until now had remained undiscovered." Bode published a table updating his concept.

Ceres's reign as a major planet, though, was a bit shorter than Pluto's. William Herschel, using his large telescope in Great Britain, was quickly able to discern that Ceres was smaller than our Moon. And Heinrich Olbers, a German physician and accomplished amateur astronomer, soon found a similar object in the same region, which he christened Pallas. Over the next five years, two more, named Juno and Vesta, were found. Being hundreds rather than thousands of miles in diameter, these newfound objects appeared starlike ("asteroidical") to Herschel in his telescope, so he suggested the name *asteroid* to describe this new class of objects. It took some time, though, for all astronomers to fully apply this term. As late as 1866, the Berlin Observatory's annual yearbook continued to list the first four asteroids as major planets. Other observatories called them "minor planets" for a while.

In the nineteenth century it was believed the asteroids were the remains of a former full-sized planet that had somehow disintegrated in the distant past. Today it is known they are a field of debris—tens of millions of fragments of planetesimals that failed to coalesce into a major planet owing to the gravitational tugs of nearby Jupiter, and that then randomly smashed into one another like cosmic bumper cars. Ceres was a protoplanet that failed to grow up.

But no tears need to be shed for this celestial goddess. At the same time that Pluto got demoted in 2006, Ceres got re-promoted. As it is the largest object in the asteroid belt (containing a third of the belt's entire mass) and rather round, with a diameter of about 590 miles (950 kilometers), the International Astronomical Union reclassified it as a dwarf planet, the sole one in the belt. It's the queen of the asteroids, majestically orbiting the Sun once every 4.6 years. NASA's *Dawn* spacecraft is currently exploring this dwarf planet, gathering data from an orbit several thousand miles above Ceres's heavily cratered surface.

CHAPTER FOUR

The Watery Allure of Mars

The renegade astronomer who started the gossip about
water on Mars

W E take them for granted when walking
along a shoreline or river bank. Looking
down, we see once-jagged rocks now
rounded and worn smooth by the flow of
running water. Pebbles have never been big news here on
terra firma—but they were an Earth-shattering, or rather a
Mars-shattering, discovery when spotted on the red planet by
the Mars rover Curiosity in 2012.

That finding and further evidence—erosion channels
carved into the Martian landscape, large expanses of sedimen-
tary deposits in former lakes (now dry craters)—all strengthen
the case that liquid water once flowed freely over the surface
of our planetary neighbor when the planet was warmer, and
its atmosphere denser, more than three billion years ago.
These recent revelations have made me wonder how Percival
Lowell would have handled the news, if he were still living
today. He's the man who infamously dominated this whole
conversation about water on Mars more than a century ago.

The oldest of five children, Lowell came from a well-established New England family. He was one of the Boston Brahmins, upper-crust Massachusetts townsmen who had made their fortunes creating the American textile industry. A few years after graduating from Harvard in 1876, Lowell traveled extensively, especially to the Far East, which led to his writing several well-received books on the region.

By the 1890s, though, restless and searching for individual expression, he renewed a childhood interest in astronomy. "After lying dormant for many years," recalled his brother, "it blazed forth again as the dominant one in his life." Independently wealthy, Lowell decided to establish his own private observatory atop a pine-forested mesa nestled against the small village of Flagstaff, Arizona (then still a territory of the United States). It was a daring venture for an amateur astronomer with no professional experience, especially since he found himself competing with the new and larger astronomical outposts then being built by universities and research institutions throughout the United States. In this rivalry, Lowell became the controversial outsider, insisting that his staff pursue the questions that interested him alone. His initial aim was to observe the particularly close approaches of Mars occurring in 1894 and 1896. Given his obsession with the red planet, the high perch on which his 24-inch (61-centimeter) refracting telescope rested more than a mile above sea level was soon dubbed Mars Hill.

Mars, with its vivid ruby luster, had been fascinating stargazers for millennia. This interest grew even more intense after the invention of the telescope. As magnifications increased over the decades, astronomers began to discern distinct markings on

Percival Lowell on the observer's chair at
Lowell Observatory's 24-inch telescope.
(Wikimedia Commons)

Mars's surface. Bright patches around its poles, similar in appearance to our own planet's Arctic and Antarctic regions, were seen to wax and wane with the Martian seasons. So earthlike were these phenomena that by 1784 the German-born British astronomer William Herschel was reporting that Mars "is not without a considerable atmosphere . . . so that its inhabitants probably enjoy a situation in many respects similar to ours."

Scrutiny of Mars was particularly favorable in the fall of 1877, when Earth and Mars were at their closest, approaching in their orbits to within 35 million miles (56 million kilometers) of one another. The superb viewing conditions allowed the Italian astronomer Giovanni Schiaparelli to map numerous dark streaks crossing Mars's reddish ochre regions. In his native language, he called these thin shadowy bands *canali*, or "channels," which many deduced arose from natural geologic processes.

But Schiaparelli's term was translated inaccurately, a gaffe that led to many fanciful conjectures. The most notorious, by far, was the assumption that the "canals" were irrigation works built by advanced beings, who were directing scarce resources over the surface of their planet for cultivation. The building of the Suez and Erie canals in the nineteenth century was still fresh in the public's memory. "Considerable variations observed in the network of waterways," wrote French astronomer Camille Flammarion in 1892, "testify that this planet is the seat of an energetic vitality. . . . There might at the same moment be thunderstorms, volcanoes, tempests, social upheavals and all kinds of struggle for life." No one championed this extravagant vision more avidly than Percival Lowell.

With the opening of his observatory in 1894, Lowell immediately began to map Mars, adding 116 waterways to Schiaparelli's original depiction. And within a year he published a book titled simply *Mars*, following up in coming years with *Mars and Its Canals* and *Mars as the Abode of Life*. The lines discerned on Mars, he declared, were assuredly artificial rivers conveying seasonal snowmelt from the planet's polar caps. That they were even visible from Earth was likely due to the massive vegetation growing along the canal banks. Promoting his ideas in books and lectures like a blue-blooded carnival

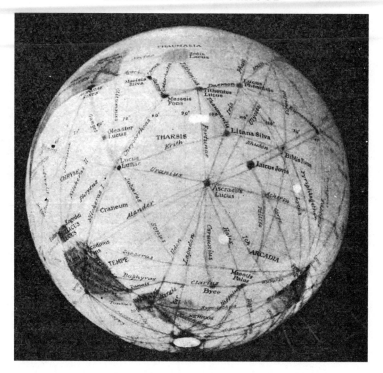

One of Percival Lowell's drawings of the Mars "canal" system.
(From *Mars as the Abode of Life*, Lowell 1908)

barker, he fancifully imagined his Martian civilization as dependent on its global irrigation system to remain extant. The very fact that the Martian features he saw were straight—like the canals, streets, and railways on Earth—increased the odds, he claimed, that they were produced by intelligent workers.

Serious astronomers were aghast at Lowell's certainty. Prestigious scientific journals refused to publish his findings. William Wallace Campbell, then director of the Lick Observatory (the Lowell Observatory's chief competitor), called

Lowell "a trial to sane astronomers." Many other observers were not seeing the same Martian features, and with good reason. "From Earth," University of New Mexico geographer K. Maria Lane has noted, "the surface of Mars was (and still is) notoriously difficult to make out. Even under excellent conditions for 'seeing,' Mars shimmered tantalizingly, allowing only fleeting glimpses of its surface." Lowell had clunkily collated his overall map from dozens of sketches of individual Martian regions, each glimpsed in a flash. A new method of planetary photography, which his observatory introduced in 1905, didn't help his case; a few dark markings were seen, but not a globe-spanning canal system.

The public and the popular press, however, reveled in Lowell's story—so much so that by 1907 the *Wall Street Journal* reported that evidence for the existence of Martian folk surpassed that year's financial panic as the news story of the year.

That media endorsement, though, was Lowell's last hurrah. Within a few years, making further observations with larger telescopes, astronomers generally concurred that Lowell's canals were merely an optical illusion—the eye imposing linearity upon an array of smaller, irregular details. The Boston Brahmin's exotic imaginings lingered long after his death in 1916 at the age of sixty-one but were finally put to rest (once and for all!) when a series of Mariner missions, launched by NASA in 1965 and 1969, showed Mars to be a completely barren world, pitted with craters.

Intriguingly, though, Mars orbiters later photographed ancient riverbeds with tributaries and erosion patterns that appeared to be carved by catastrophic flooding episodes. Probes now roaming over the Martian landscape confirmed those observations. Perhaps Lowell would have been elated that there were Martian channels after all. But these passages were forged by water flowing naturally, and in Mars's distant past rather than in the present day. In the end, to Lowell's likely dismay, there were no little green men digging trenches.

CHAPTER FIVE

Rings, Rings, Rings

Finding that planets could have rings

To see this amazing celestial feature in person, hop a spaceship and rocket toward the Scorpius and Centaurus constellations. After traveling a distance of some four hundred light-years, you'll come upon an astounding sight—a ringed planet that makes Saturn's rings look scrawny by comparison.

Several years ago, astronomers had observed this planet's sun, known simply as J1407, undergo a complex series of eclipses. Over the course of fifty-six days, the star's light brightened and dimmed erratically. What could be causing such fluctuations? Astronomers from both the University of Rochester and Leiden Observatory in the Netherlands have suggested that those repeated eclipses were due to the transit of a giant ringed planet orbiting the star.

And not just any ringed planet. According to their model, this exoplanet's rings extend outward for some 56,000,000 miles (90,000,000 kilometers). Such a disk would be quite a sight if it resided in our solar neighborhood, as its radius is more

What J1407b's ring system would look like
at dusk in the skies above Leiden University in the
Netherlands, if it were in Saturn's orbit.
(M. Kenworthy/Leiden)

than half the distance from the Sun to the Earth. Saturn's most
prominent rings reach out a mere 175,000 miles (282,000 kilometers) from the planet's equator. This colossal ring system is
one of the first suspected to reside outside our solar system. And
its discovery was announced nearly four centuries after Saturn's
planetary hula hoop was first recognized for what it was.

As with so many seminal moments in astronomy, the
long path toward understanding that a planet could even be

surrounded by a ring began with Galileo. With his publication of *Sidereus nuncius,* the "Starry Messenger," in March 1610, Galileo first announced to the world the cosmos-shattering revelations spied through his homemade telescope: that the lunar landscape was filled with mountains and craters; that a multitude of stars blended together to form the Milky Way's luminous white band; and that the planet Jupiter, like some mini-solar system, was repeatedly circled by a set of moons.

But that was just the start. Four months later, once Saturn became visible in the nighttime sky, Galileo turned his telescope to what was then the farthest known planet. And what he encountered he called a "very strange wonder." While keeping his discovery secret from fellow scientists for several months, Galileo swiftly notified the secretary of his Medici patron, the Grand Duke of Tuscany. "The star of Saturn is not a single star," disclosed Galileo, "but is a composite of three, which almost touch each other."

With the poor quality of his rudimentary telescope, Galileo was, of course, erroneously seeing Saturn's ring system as two small blobs, perched on either side of the bigger central planet. The seventeenth-century Venetian poet Giulio Strozzi, in an ode to the great astronomer, lyrically described the sight as "in three minor knots divided."

Likely thinking of Saturn's appendages as separate moons, much like Jupiter's, Galileo aimed to keep track of how they orbited the planet. But, to Galileo's great surprise, Saturn's telescopic image instead underwent "a strange metamorphosis," changing back and forth over the years. Johann Locher, an astronomy student in Bavaria, made this cyclic transformation the subject of his dissertation in 1614. "Saturn deceives or really mocks the astronomers out of hatred or malice. For [the planet] has projected various appearances," he wrote. "Sometimes he is seen single and sometimes triple; at one time elongated and at other times round." By 1616, Saturn looked as if it had handles. All these variations were due to how Saturn's rings were positioned with respect to the Earth, although astronomers didn't know that yet.

By 1650, according to astronomy historian Albert Van Helden, "the problem of Saturn's appearances had become a celebrated puzzle." Astronomers were wondering whether Saturn was round, egg-shaped, or composed of three bodies.

It's easy to assume that better telescopes eventually solved the mystery, but that wasn't fully the case. There was also some clever thinking involved. The inventive Dutch astronomer Christiaan Huygens had built a fifty-powered telescope that allowed him in 1655 to discover Saturn's first moon, Titan. Saturn itself, as Huygens described it, then had "arms extended on both sides in a straight line, as though the planet were pierced through the middle by a kind of axis." By the start of 1656 these arms had vanished altogether. Despite this disappearing act, Huygens still reasoned that Saturn's chameleonic changes could be explained by the planet being "surrounded by a thin flat ring, nowhere touching, and inclined to the ecliptic." First keeping this knowledge secret, needing more time to flesh out his theory and observe the ring with an even better telescope, Huygens finally made it public in his *Systema Saturnium*, published in 1659.

His fellow astronomers, however, did not greet the ring hypothesis with open arms. An accomplished observer in Rome, Honoré Fabri, declared it "pure fiction." He preferred to think that Saturn was merely accompanied by several satellites. But within a decade, as telescopes improved, even Huygens's harshest critics came to accept his explanation.

From the start Huygens imagined the ring as solid, like some kind of celestial phonograph record. But that assumption was considerably undermined in 1675 when Giovanni Cassini, director of the Paris Observatory, discovered that Saturn's ring had a prominent gap, now known as the Cassini division. Cassini suspected that the ring was composed of small celestial bodies, a notion spurned by most astronomers. But a century later, the French mathematician Pierre-Simon Laplace offered a further argument against the solid-ring idea. He demonstrated mathematically that a solid structure would be highly unstable.

It was not until the nineteenth century that both theory and observation at last resolved the makeup of Saturn's rings once and for all. In a prize-winning 1856 essay, the Scottish physicist James Clerk Maxwell (who several years later went on to develop his historic theory of electromagnetism) lucidly proved that the ring had to be composed of innumerable particles, each orbiting Saturn like a minuscule moon. It was the only configuration that remained durable against gravitational and centrifugal forces. All doubts were erased in 1895 when James E. Keeler, then director of the Allegheny Observatory in Pennsylvania, pegged the velocity of Saturn's rings. Newton's law of gravity predicted that the tiny chunks circulating in the outer part of the ring would travel more slowly than those closer in—just as Neptune, far from the Sun, orbits at a lower velocity than the solar system's inner planets. And that's exactly what Keeler measured. Within days of his observation, he sent a report to the *Astrophysical Journal* ("A Spectroscopic Proof of the Meteoric Constitution of Saturn's Rings"), triggering a torrent of magazine and newspaper articles around the world.

Saturn's ring material, composed largely of ice and dust, ranges in size from grains to boulders the size of a house and larger. This material may have originated when an ancient ice-cloaked Saturnian moon was either ripped apart by tidal forces or shattered by an incoming comet. Or possibly it is simply material left over from the nebular disk out of which Saturn itself formed.

Saturn lost its special status as our solar system's sole ringed planet in the 1970s and 1980s, when both telescopic observations and spacecraft flybys of the other gas giants—Jupiter, Uranus, and Neptune—spotted rings around them as well. It took longer to find these ring systems, as they are far less substantial, hence fainter and difficult to see.

That wouldn't be the case for Saturn Giganticus, or J1407b, as the exoplanet is officially known. If it replaced Saturn within our solar system, the rings would appear many times larger than the width of the full Moon, and with our eyes alone we'd be able to marvel at their beauty during a long, dark night.

CHAPTER SIX

The Baffling White Dwarf Star

Discovering this star opened up a whole
can of cosmic worms

Iｎ 1862 the first hint arrived that the stellar universe
was far stranger than anyone imagined—or could
imagine. It came with the knowledge that a faint companion slowly circles Sirius, the brightest star in the
nighttime sky.

Astronomers at the time didn't recognize what they had
uncovered. It would take decades—until the 1910s—for them
to fully realize that Sirius B, as the tiny companion came to be
known, was a star like no other seen before. Once its nature
was revealed, though, it didn't take long for theorists to con-
ceive of other bizarre creatures that might be residing in the
stellar zoo.

The story begins, not in 1862, but actually two decades
earlier. For a number of years, the noted German astronomer
Friedrich Wilhelm Bessel, director of the Königsberg Observatory, had been going through old stellar catalogs, as well as
making his own measurements, to track how the stars Sirius
and Procyon were moving across the celestial sky over time.

By 1844 he had enough data to announce that Sirius and Procyon were not traveling smoothly, as expected; instead, each star displayed a slight but distinct wobble—up and down, up and down. With great cleverness, Bessel deduced that each star's quivering walk meant it was being pulled on by a dark, invisible companion circling it. Sirius's unseen companion, he estimated, completed one orbit every fifty years.

Bessel was clearly excited by his find; in his communication to Great Britain's Royal Astronomical Society, he wrote, "The subject . . . seems to me so important for the whole of practical astronomy, that I think it worthy of having your attention directed to it."

Astronomers did take notice, and some tried to discern Sirius's companion through their telescopes. Unfortunately, at the time Bessel reported his discovery, Sirius B was at its closest to gleaming Sirius, from the point of view of an observer on Earth, and thus lost in the glare. But even years later, no one was successful in spotting the companion.

That all changed on January 31, 1862. That night in Cambridgeport, Massachusetts, Alvan Clark, the best telescope manufacturer in the United States, and his younger son, Alvan Graham Clark, were testing the optics for a new refractor they had been building for the University of Mississippi. It was going to be the biggest refracting telescope in the world. Looking at notable stars to carry out a color test of their 18.5-inch (47-centimeter) lens, the son observed a faint star very close to Sirius.

This momentous sighting might have gone unrecorded. But fortunately, the father was an avid double-star observer and possibly encouraged his son to report the discovery to the nearby Harvard College Observatory. In fact, according to historian Barbara Welther, rather than its being an accidental discovery, as long asserted in astronomy books, "there might have been a [prearranged] connection between the elder Clark and someone at Harvard" to look for Sirius's companion.

Whatever the case, George Bond, the observatory's director, confirmed the find a week later, and he soon wrote up two papers, one submitted to a German journal of astronomy, the other to the *American Journal of Science*. One question was uppermost on Bond's mind: "It remains to be seen," he wrote, "whether this will prove to be the hitherto invisible body disturbing the motions of Sirius." The newfound star seemed to be in the right place to explain the direction of Sirius's wavelike motions, but its luminosity was extremely feeble—so dim, in fact, that it suggested at the time a star too small to have enough mass to account for the wobble. Here was the first clue to Sirius B's uniqueness. For revealing Sirius's dark companion, Alvan Graham Clark in 1862 garnered the prestigious Lalande Prize, presented by the French Academy of Sciences for the year's most outstanding achievement.

As astronomers around the globe continued over the years to observe the orbital dance of Sirius and its partner, they eventually determined that the companion was hefty enough (a solar mass) to pull on Sirius, though with a light output less than a hundredth of our Sun's. But no one worried about this disparity at first. They just figured it was a sunlike star cooling off at the end of its life. At this point, no one had yet secured a spectrum of the light emanating from Sirius B, a difficult task owing to the overwhelming brightness of the binary's primary star. Astronomers just assumed it had to be yellow or red, like other dim and cooler stars. Astronomy had a general rule at the time: the hotter the star, the brighter. The brightest stars' colors were white, blue-white, or blue.

But in 1910, Princeton astronomer Henry Norris Russell noticed something in a past observation that cast doubt on that rule. On a Harvard College Observatory photographic plate, a faint companion of the star 40 Eridani—a companion known since 1789—was labeled as blue-white. How could that be? Russell doubted that such a classification could be correct for such a faint star. But in 1914, Walter Adams at the Mount Wilson Observatory in California confirmed the spectrum. The star was indeed white-hot, yet dim. "I was flabbergasted," recalled Russell. "I was really baffled trying to make out what it meant." Then, in 1915, Adams determined that Sirius's faint companion, too, displayed the spectral features of a blazing blue-white star.

Soon, theorists, such as the British astrophysicist Arthur S. Eddington, figured out what was going on. If a star is both white and hotter than our Sun, it must be emitting more light over each square inch of its surface. But since Sirius B is so faint, that could only mean it had less surface area than our Sun—in other words, it is far smaller, roughly the size of the Earth. Such stars came to be called "white dwarfs."

But how does a Sun's worth of mass get squeezed into such a tiny volume? As Eddington later remarked mischievously, "The message of the companion of Sirius when it was decoded ran: 'I am composed of material 3,000 times denser than anything you have ever come across; a ton of my material would be a little nugget that you could put in a matchbox.' What reply can one make to such a message? The

An artist's impression of the blue-white star Sirius A (left) and its
tiny white-dwarf companion Sirius B (right) as they might appear
to an interstellar visitor.
(NASA/ESA/G. Bacon [STSci])

reply which most of us made . . . was—'Shut up. Don't talk
nonsense.'"

It took quantum mechanics, under development in the
1920s, to solve the puzzle. By 1926 British theorist Ralph
Fowler finally figured out that the density inside the compact
dwarf star becomes so extreme that all its atomic nuclei, like
droves of little marbles, are packed into the smallest volume
possible, while its free electrons generate an internal energy
and pressure that keeps it from collapsing even further. This
creates an ultraconcentrated material impossible to assemble
on Earth. Astronomers later learned that this is the end stage

for a star like our Sun. The white dwarf is the luminous stellar core left behind after the star runs out of fuel and releases its gaseous outer envelope into space. Such will be our Sun's fate some five billion years from now.

The discovery of the extremely dense white dwarf star turned out to be only the first volley in a startling stellar revolution. By the 1930s, working with the new laws of both quantum mechanics and relativity, theorists were astonished (and disturbed) to find that dying stars might face even stranger fates, if they had enough mass. Discovery of the white dwarf had opened up a whole can of cosmic worms.

In the early 1930s a young man from India named Subrahmanyan Chandrasekhar, while about to start graduate work at Cambridge University, calculated that if the mass of a white dwarf passes beyond a certain limit (now known to be 1.4 solar masses; that is, 1.4 times the mass of our Sun), it will collapse, its radius approaching zero as the star is overcome by the extreme pressure of gravity. What happens to the star? Chandrasekhar didn't know. All he could say for sure was that a "star of large mass . . . cannot pass into the white-dwarf stage, and one is left speculating on other possibilities."

The great Eddington declared that "there should be a law of nature to prevent a star from behaving in this absurd way!" But, with the discovery of a new atomic particle—the neutron—in 1932, others ventured that the star might end up as a relatively tiny ball of neutrons, not much wider than a city.

J. Robert Oppenheimer, who went on to become the father of the atom bomb, briefly dabbled in the subject, joining with two of his graduate students to ponder a neutron star's range of stable masses. And in these deliberations he and another student, Hartland Snyder, in 1939 calculated that past a certain threshold of mass, the neutron star itself would not endure but instead face "continued gravitational contraction." The neutrons could no longer serve as an adequate brake against collapse. Oppenheimer and Snyder found that the last light waves to flee get so drawn out by the enormous pull of gravity that the rays become invisible, and the star vanishes from sight. The star literally closes itself off from the rest of the universe. "Only its gravitational field persists," reported Oppenheimer and Snyder. By 1968, astronomers regularly began calling these objects "black holes."

Today astronomers recognize that galaxies are peppered with both black holes and rapidly spinning neutron stars (we know them as pulsars). And our understanding of such zany stellar outcomes commenced, in a way, with the discovery of Sirius's faint companion, first spotted (maybe by accident, maybe not) more than 150 years ago.

CHAPTER SEVEN

The Star No Bigger Than a City

"Look happy dear, you've just made a Discovery!"

I N the fall of 1967, the first neutron star was detected, a discovery that came as a complete surprise to one and all. While the existence of such a compact star—a mere dozen miles wide—was not unforeseen (as pointed out in the previous chapter), no one imagined it would be emitting clocklike radio pulses. "No event in radio astronomy seemed more astonishing and more nearly approaching science fiction," said the British radio astronomy pioneer James S. Hey. And it was a long road to that flabbergasting finding.

Subrahmanyan Chandrasekhar, while starting his research career at Cambridge University before moving to the United States, spent several years in the early 1930s trying to convince his colleagues in the British astrophysics community that if a star were massive enough it would never settle down as a white dwarf star in its old age. Instead, his calculations indicated that the dwarf would undergo further stellar collapse. While Chandra (as he was best known) never speculated on the other forms the star might take, others boldly did.

At a 1933 meeting of the American Physical Society, Walter Baade of the Mount Wilson Observatory in California and Fritz Zwicky at Caltech introduced the idea that such a massive sun might end up as a neutron star, a dense ball of packed neutrons not much wider than a city. This transformation would occur, they reported, in a spectacular stellar explosion they had christened a "supernova."

Astronomers had long recognized that novae—"new stars"—occasionally appeared in the heavens. By the early twentieth century, they realized that this phenomenon involved some kind of outburst on the star. Moreover, they began to notice that there were two kinds. There were the "common" novae that appeared up to thirty times a year in both the Milky Way and other galaxies (now known to occur when a white dwarf steals mass from a companion—matter that compresses on the dwarf and eventually ignites in a thermonuclear blast). And then there was a special set, far more luminous and much rarer. In his native German, Baade first referred to them as *Hauptnovae* (chief novae). But both Zwicky and Baade translated that into English as "supernovae" during their lectures in Pasadena.

More than providing a name, Zwicky and Baade offered a reason for the spectacular flare-up. Neutrons had just been discovered by particle physicists in 1932, and even before that the Soviet physicist Lev Landau had suggested that the compressed cores of massive stars might be "forming one gigantic nucleus," as he put it. Zwicky and Baade took the idea further by suggesting that under the most extreme conditions—during the explosion of a star—suns would transform completely into naked spheres of neutrons. The stellar core would somehow implode, pressing together all its positively charged protons and negatively charged electrons to form a compact ball of neutral particles.

This proposal was considered wildly speculative, and only a handful of physicists, including J. Robert Oppenheimer and his student George Volkoff, proceeded to investigate a neutron star's possible structure, recognizing how nuclear forces would keep such stars from further collapse. For some three decades, neutron stars remained only theoretical inventions, which astronomers figured would never be seen even if they did exist, due to their extremely small size. Even the notable Princeton theorist John Archibald Wheeler was shortsighted at first. In 1964 he published an article on the neutron star, in which he said, "There is about as little hope of seeing such a faint object as there is of seeing a planet belonging to another star." But Wheeler's prediction was soon thwarted in a mere three years—thanks to a bit of serendipity.

A small platoon of students and technicians, led by Cambridge University radio astronomer Antony Hewish, had just completed the construction of a sprawling radio telescope near the university: more than two thousand dipole antennas, lined up like rows of corn and connected by dozens of miles of wire. Jocelyn Bell, a native of Ireland, was one of the laborers: "I like to say that I got my thesis with sledgehammering," she once joked.

The telescope was designed to passively search for fast variations in the intensities of pointlike radio sources, such as quasars, while the celestial sky moved overhead. The data continually registered on a strip-chart recorder, and it was Bell's job to analyze the long stream of paper—ninety-six feet

Jocelyn Bell helped build this radio telescope
that discovered the first pulsar in 1967.
(Graham Woan)

(29 meters) each day—for her doctoral dissertation. Upon
reviewing the first few hundred feet, she noticed, "There was
a little bit of what I call 'scruff,' which didn't look exactly like
[manmade] interference and didn't look exactly like [quasar]
scintillation. . . . I began to remember that I had seen some of
this unclassifiable scruff before, and what's more, I had seen it
from the same patch of sky."

Eventually observing it with a higher-speed recording,
Bell (later Bell Burnell upon marriage) came to see that the
scruff was actually a methodical succession of pulses spaced
1.3 seconds apart. The unprecedented precision caused Hewish and his group to briefly label the source LGM, for "Little
Green Men." This was done only half in jest. At one point,

some consideration was given to the possibility that the regular pulsations might be coming from an extraterrestrial-built beacon, which annoyed Bell a bit: "I was [then] two-and-a-half years through a three-year studentship and here was some silly lot of Little Green Men using *my* telescope and *my* frequency to signal the planet Earth."

But within a few months, Bell uncovered three more rhythmical signals in different regions of the sky (along with getting engaged to be married between the second and the third). There was no more mention of outer-space aliens. It was highly unlikely, she said, that there were "lots of little green men on opposite sides of the universe" using the same frequency to get Earth's attention. Carefully kept under wraps, the news was finally released in February 1968, and upon discovering a pretty, young woman was involved, the press went wild. "One of [the photographers] even had me running down the bank waving my arms in the air—Look happy dear, you've just made a Discovery!" Inspired by the name of the recently discovered quasars, a British science journalist dubbed the novel objects *pulsars*, for pulsating stars, a label that astronomers swiftly adopted.

In their *Nature* report, Hewish, Bell, and three colleagues pointed out that the exceedingly short span of the beep itself—around a hundredth of a second—meant that the source could span no more than 5,000 kilometers (3,100 miles, around the distance light can travel in a hundredth of a second, close to the width of the planet Mercury). This suggested the pulsar was either a white dwarf or neutron star.

The Cambridge team at first wondered whether the entire star was pulsating in and out, with the radiation then "likened to radio bursts from a solar flare occurring over the entire star

Jocelyn Bell in 1967, at the time she
was working on revealing
the first neutron star.
(Roger Haworth, Wikimedia Commons)

during each cycle of the oscillation." Within months, though, Cornell University theorist Thomas Gold developed the model that best explained a pulsar's behavior: it was most likely a neutron star, whose highly magnetized body as it rapidly spins transfers the rotational energy into electromagnetic energy. This radiation is then beamed outward like a lighthouse beacon from its north and south magnetic poles. Depending on the pulsar's alignment with Earth, we observe either one or two blips of radio energy with each pulsar rotation.

Since neutron stars can spin quite fast, Gold predicted that radio astronomers should also detect pulsars with shorter periods than those first discovered. This was successfully confirmed when astronomers found extremely fast-spinning pulsars within the Vela and Crab Nebulae—with periods of 0.089 and 0.033 seconds, respectively. Since each nebula was a supernova remnant, these finds also validated Zwicky and Baade's original assertion that neutron stars would be found at the sites of stellar explosions. You can think of a pulsar as a stellar tombstone, which marks the spot where a giant star, too heavy to die quietly as a white dwarf, tore itself apart in a brilliant explosion.

Zwicky had imagined that the stellar explosion somehow created the neutron star. But astronomers later realized it was the other way around. Once it runs out of nuclear fuel, the massive star's core collapses catastrophically under the force of gravity. A core that was once the size of the Moon is squeezed down *in less than a second*, cramming the mass of 1.4 to 2.5 suns into a space roughly as wide as Philadelphia. In this way the stellar protons and electrons merge to form a tight ball of neutrons, whose density is so great that a sugar-cube-sized portion would weigh as much as Mount Everest. The shock wave sent out from the collapse, along with a flood of neutrinos, then speeds through the remaining stellar envelope, emerging from the surface as the spectacular supernova.

Astronomers estimate that at least a few hundred million neutron stars now reside in the Milky Way, created over the eons since our galaxy's birth. But the first one revealed, officially known as PSR 1919+21 for its celestial coordinates, will never be forgotten—not just for its discovery but for the controversy that later surrounded it. When the Nobel Prize in Physics was awarded in 1974 for pioneering work in radio astrophysics, including the discovery of pulsars, it was Hewish who walked up to the podium (along with Martin Ryle), but not Bell Burnell. Hewish had been skeptical about Bell's "scruff" at first, believing at one point that it was either a stellar flare or manmade. It was only due to her persistence that its origin was at last revealed. At Great Britain's *Observatory* magazine, the editors wryly joked among themselves that Nobel now stood for "No Bell." Her being a young, female graduate student (only two women have won the physics prize since the first award ceremony in 1901) likely prejudiced the judges.

But Bell Burnell, who went on to a distinguished career as a professor, dean of science, and president of the Royal Astronomical Society, maintained a sanguine attitude about this flagrant oversight. During an after-dinner speech at a relativity conference in 1977, she noted that the final responsibility for the success or failure of a scientific project rests with its supervisor. "I believe it would demean Nobel Prizes if they were awarded to research students, except in very exceptional cases, and I do not believe this is one of them. . . . I am not myself upset about it—after all, I am in good company, am I not!"

No—they're in good company with *her*. While memories of who won a Nobel Prize dim over time, Bell Burnell will always serve as the main protagonist when recounting the story of the neutron star's discovery.

CHAPTER EIGHT

Ye Old Black Hole

An eighteenth-century theorist was just
too far ahead of his time

BORN on Christmas day in 1724, the Englishman
John Michell was a geologist, astronomer, mathematician, and theorist who regularly hobnobbed
with the greats of the Royal Society of London. His
companions included such men as Henry Cavendish,
Joseph Priestley, and even the Society's American fellow
Benjamin Franklin (during the diplomat's two long stays in
London). The claim could be made, science historian Russell
McCormmach has written, that Michell was "the most inventive of the eighteenth-century natural philosophers." Yet until
recently, if he was remembered at all, it was for his suggestion,
in 1760, that earthquakes propagate as elastic waves through
the Earth's crust. That earned Michell the title "father of
modern seismology." In addition, a torsion balance he invented was later used by Cavendish to weigh the entire Earth.

Otherwise, Michell has been largely forgotten. That's
because he had the unfortunate habit of burying original
insights—such as the inverse-square law of magnetic force—

in journal papers that focused on inferior research. Some of his greatest ideas were casually mentioned in brief asides or footnotes. As a consequence, long-lasting fame eluded him.

Michell began his scientific investigations at Queens' College in Cambridge. Son of an Anglican rector, he entered Queens' in 1742 at the age of seventeen and after graduation remained there to teach for many years, eventually becoming a rector as well. A contemporary described him as a "short Man, of a black Complexion, and fat. . . . He was esteemed a very ingenious Man, and an excellent Philosopher."

But by 1763, ready to marry, Michell decided to devote himself to the church. He ultimately settled in the village of Thornhill in West Yorkshire, where he served as a clergyman until his death in 1793 at the age of sixty-eight. Yet, over those years with the Church of England, the reverend continued to indulge his wide-ranging curiosity. He had a nose for interesting questions and was willing to stick his neck out in speculation, though always grounded in his first-rate mathematical skills. One of Michell's more intriguing conjectures at this time, right when Great Britain was recovering from its war with colonial America, was imagining what today we would call a black hole.

This idea grew out of an earlier prediction that Michell had made. Astronomers in the eighteenth century were starting to see more and more double stars as they scanned the celestial sky with their ever-improving telescopes. The common wisdom of the time declared that such stars were actually at various distances from Earth and closely aligned in the sky by chance alone—that it was just an illusion that they were connected in any way. But, with remarkable insight,

Michell argued that nearly all those doubles had to be gravitationally bound together.

He was suggesting that some stars exist in pairs, a completely novel notion. In a groundbreaking paper published in 1767, he worked out the high probability that, given how most other stars were arranged in the sky, the twin stars were physically near one another—"the odds against the contrary opinion," he stressed, "being many million millions to one." (As usual, he displayed the results in a footnote.) In carrying out this calculation, Michell was the first person to add statistics to astronomy's repertoire of mathematical tools. The paper was "arguably the most innovative and perceptive contribution to stellar astronomy . . . in the eighteenth century," according to the astronomy historian Michael Hoskin.

At the same time, Michell recognized that double stars would be quite handy for learning lots of good things about the properties of stars—how bright they are, how much they weigh, how vast is their girth. Two stars orbiting each other were the perfect laboratory for testing out Newton's laws of gravity from afar and arriving at answers. Yet, nearly all astronomers in his day weren't concerned with such questions. They were too busy discovering new moons or tracking the motions of the planets with exquisite precision. To them, the stars were merely a convenient backdrop for their measurements of the solar system, the arena that most captured their attention.

The British astronomer William Herschel, a friend of Michell's, was the rare exception to that emphasis, and within a dozen years of Michell's paper on double stars, he began monitoring and cataloging the stars positioned close together in the sky. Encouraged by Herschel's growing data bank,

Michell decided to extend his ideas on double stars in a paper with the marathonic title "On the Means of discovering the Distance, Magnitude, &c. of the Fixed Stars, in consequence of the Diminution of the Velocity of their Light, in case such a Diminution should be found to take place in any of them, and such other Data should be procured from Observations, as would be farther necessary for that Purpose." (*Whew!*) It was in this work that Michell hinted at the possibility of a black hole—or at least his eighteenth-century, Newtonian version of one.

The eminent Henry Cavendish, discoverer of hydrogen and its connection to water, read Michell's paper before the Royal Society over three successive meetings in December 1783 and January 1784. It was then published in the Royal Society's *Philosophical Transactions*, taking up twenty-three pages in print. Michell was devoted to the Society and at least once a year traveled the arduous two hundred miles (three hundred kilometers) from Yorkshire to London to either attend its meetings or meet with Society friends. But for those December and January meetings, the reverend inexplicably stayed home. It could have been ill health, but some historians have speculated that Michell recognized the daring nature of his paper and thought it would be more readily accepted if his close friend and highly respected colleague presented it to the Society.

The radical technique that Michell was proposing to apply to study the stars involved the speed of light. If astronomers closely monitored the two stars in a binary system moving around each other over the years, noted Michell, they could calculate the masses of the stars. It was a basic application of Newton's laws of gravity. And if the motions of paired stars

VII. *On the Means of discovering the Distance, Magnitude,* &c. *of the Fixed Stars, in consequence of the Diminution of the Velocity of their Light, in case such a Diminution should be found to take place in any of them, and such other Data should be procured from Observations, as would be farther necessary for that Purpose. By the Rev.* John Michell, *B. D. F. R. S. In a Letter to* Henry Cavendish, *Esq. F. R. S. and A. S.*

Read November 27, 1783.

DEAR SIR,   Thornhill, May 26, 1783.

THE method, which I mentioned to you when I was last in London, by which it might perhaps be possible to find the distance, magnitude, and weight of some of the fixed stars, by means of the diminution of the velocity of their light, occurred to me soon after I wrote what is mentioned by Dr. PRIESTLEY in his History of Optics, concerning the diminution of the velocity of light in consequence of the attraction of the sun; but the extreme difficulty, and perhaps impossibility, of procuring the other data necessary for this purpose appeared to me to be such objections against the scheme, when I first thought of it, that I gave it then no farther consideration. As some late observations, however, begin to give us a little more chance of procuring some at least of these data, I thought it would not be amiss, that astronomers should be apprized of the method, I propose (which, as far as I know,

F 2   has

The eighteenth-century scientific paper in which John Michell first suggested the existence of the Newtonian version of a "black hole."
(*Philosophical Transactions of the Royal Society of London*, Michell 1784)

were affected by each other's gravitation, suggested Michell, light should also be affected. This was an era when light was assumed to be made up of "corpuscles," swarms of particles—largely because the great Newton had championed that idea.

Now imagine those particles journeying off a star and out into space. Michell assumed that they, too, would be attracted by gravity, just as matter is. The more sizable the star, the stronger the gravitational hold upon the light, slowing down its speed. There would be a "diminution of the velocity of [the stars'] light," as the title of his paper announced. Measure the velocity of a beam of starlight entering a telescope and, voilà, you obtain a means of weighing the star.

This is where the "black hole" possibility arises: Michell took this scenario to the extreme and estimated when the mass of the star would be so great that "all light . . . would be made to return towards it, by its own proper gravity"—like a spray of water shooting up from a fountain, reaching a maximum height, and then plunging back down to the bowl. With not one radiant corpuscle escaping from the star, it would remain forever invisible, like a dark pinpoint in the sky. According to Michell's calculations, this transformation would occur when the star was about five hundred times wider than our Sun and just as dense throughout. In our solar system, such a star would extend past the orbit of Mars.

In 1796, in the midst of the French Revolution, the mathematician Pierre-Simon Laplace independently arrived at a similar conclusion. He briefly mentioned these *corps obscurs*, or hidden bodies, in his famous *Exposition du Système du Monde* (The system of the world), essentially a handbook on the cosmology of his day. "A luminous star of the same density as the

Earth, and whose diameter should be two hundred and fifty times larger than that of the Sun," he wrote, "would not, in consequence of its attraction, allow any of its ray to arrive at us; it is therefore possible that the largest luminous bodies in the universe, may, through this cause, be invisible." Laplace's estimate for the width of the dark star differed from Michell's because he assumed a greater density for sunlike bodies.

But did it even make sense to predict the existence of stars that could never be seen? Laplace may have had second thoughts, or simply a loss of interest. In subsequent editions of *Système du Monde*, which he published up until his death in 1827, he expunged his invisible-star speculation and never referred to it again. Michell, on the other hand, displayed greater ingenuity by suggesting a way to "see" such invisible stars. If one of them revolved around a luminous star, he noted, its gravitational effect upon the bright star's motions would be noticeable. It's the very way that astronomers today track down black holes.

In the end, though, Michell and Laplace were getting ahead of themselves, contemplating problems before the physics was in place to answer them. They didn't yet realize that supergiant stars have far lower densities than the ones they envisioned. They also never considered that the same invisibility effect could happen if a star were smaller but very, very dense. They just assumed that all stars shared the same density as the Sun or Earth. Could anything be more dense than the elements found on Earth? It seemed unthinkable in the late eighteenth century.

Both Michell and Laplace were working with an inadequate law of gravity and the wrong theory of light. They didn't yet know that light never slows down in empty space. Proving the existence of such dark stars required more advanced theories of light, gravity, and matter. The modern conception of the black hole would not emerge for nearly a century. It had to wait for the entrance of the twentieth century's most inventive natural philosopher, Albert Einstein.

CHAPTER NINE

As Though No Other Name Ever Existed

How the term "black hole" entered the scientific
literature

T HE term "black hole" has a deep, dark past and a notorious reputation. In June 1756, on the banks of the Hooghly River in Calcutta, India, at the British garrison of Fort William, 144 British men and two women were taken prisoner by the troops of the new Nawab of Bengal, Siraj-ud-daula. Siraj's men incarcerated at least sixty-four of the hostages for a night in a tiny cell known as the "black hole."

Only twenty-one survived the hot night, which was suffocating—literally. Ever since that horrific event, the words "black hole" have referred to a place of confinement, a locked cell, where it was anticipated that once you went in, you never came out. How did the term come to signify objects in outer space?

Toward the end of the 1960s, when astronomers were coming to recognize that massive stars upon running out of fuel might actually collapse to a singular point (with, theoretically, infinite density and zero volume), they had a problem. For many years, theorists had been calling such an entity a "gravitationally collapsed object," a real mouthful to pronounce over and over again in a lecture. Soon, some shortened the awkward phrase to "collapsar," while others preferred "dark star." In short, the terminology kept shifting. That all changed in 1967, when the noted Princeton University physicist John Archibald Wheeler supposedly linked the term "black hole" to the cosmos. The attribution of that lexical connection, however, has recently been challenged.

Physicist John Wheeler was often credited
with assigning the phrase "black hole" to
gravitationally collapsed stellar objects.
(Photograph by Roy Bishop, Acadia University,
courtesy AIP Emilio Segrè Visual Archives)

Wheeler usually told his side of the story in the following fashion. It was the fall of 1967, and he was attending an impromptu conference at the NASA Goddard Institute for Space Studies in New York City. Pulsars had just been detected for the first time, and astronomers were asking whether those mysterious, pulsed radio waves were coming from red giant stars, white dwarfs, or neutron stars. According to Wheeler, he told the assembly that his "gravitationally collapsed objects" might be responsible. "Well, after I used that phrase four or five times, somebody in the audience said, 'Why don't you call it a black hole?' So I adopted that," said Wheeler.

While pulsars were first detected in 1967, however, their existence remained a well-kept secret until 1968; the public announcement of the discovery was made in February 1968, when a paper on the topic was finally published in the journal *Nature*. Did Wheeler misremember the nature of his fall conference? There was a meeting on supernovae at the Goddard Institute in November 1967, but Wheeler's name is missing from the official conference proceedings.

What Wheeler did do, without dispute, is use the phrase "black hole" during an after-dinner talk at the annual meeting of the American Association for the Advancement of Science in New York City on December 29, 1967. The term then made it into print when an article based on his talk, titled "Our Universe: The Known and the Unknown," was published in *American Scientist* in 1968. Wheeler's enduring credit for introducing the phrase is due to that popular paper.

Yet firm evidence exists that the term actually arose much earlier, even in print. It was casually bandied about at the 1963 Texas Symposium for Relativistic Astrophysics. Reporting on the Texas conference, the science editor for *Life* magazine at the time, Albert Rosenfeld, used the term "black hole" in an article on the newly discovered quasars. He noted how astrophysicists Fred Hoyle and William Fowler suggested that the gravitational collapse of a star might explain the quasar's energy. "Gravitational collapse would result in an invisible 'black hole' in the universe," wrote Rosenfeld. Rosenfeld today is sure he didn't invent the term but overheard it at the meeting, although he cannot recall the source.

The phrase was mentioned again a week later at an American Association for the Advancement of Science meeting held in Cleveland. Ann Ewing of *Science News Letter* reported that astronomers and physicists at the conference were suggesting that "space may be peppered with 'black holes.'" The person who used the term there was Goddard Institute physicist Hong-Yee Chiu, who had originated the term *quasar* in 1964 in *Physics Today* and had also attended the Texas Symposium. Was he introducing another fun term to the public? No, answers Chiu; he borrowed it from the man who may have coined the phrase from the start.

From 1959 until 1961 Chiu was a member of the Institute for Advanced Study in Princeton, and during that time Princeton physicist Robert H. Dicke, both an experimental and a theoretical contributor to the study of gravitation, spoke at a colloquium about how general relativity predicted the complete collapse of certain stars, creating an environment where gravity was so strong that neither matter nor light could escape. "To the astonished audience, he jokingly added it was like the 'Black Hole of Calcutta,'" recalls Chiu. A couple of years later, when Chiu started working at the Goddard Institute, he

Physicist Robert Dicke was likely the first to
introduce the term *black hole*, during a lecture at
Princeton in the early 1960s.
(Courtesy AIP Emilio Segrè Visual Archives, *Physics Today*
Collection)

heard Dicke there casually use the phrase once again during a
series of visiting lectures. In this way, Dicke may have released
the term into the scientific atmosphere.

Loyola University physicist Martin P. McHugh, while
working on a biography of Dicke, discovered it was one of
Dicke's favorite expressions. He often used it with his family
in an entirely different context. His sons told McHugh that
their father exclaimed, "Black Hole of Calcutta!" whenever a
household item appeared to have been swallowed up and
gone missing.

Yet, Wheeler still deserves a large portion of the credit for placing the phrase into the scientific lexicon. Given Wheeler's status in the field, his decision to adopt the moniker bestowed a gravitas upon it, giving the science community permission to embrace the term without embarrassment. "He simply started to use the name as though no other name had ever existed, as though everyone had already agreed that this was the right name," said his former student, Caltech physicist Kip S. Thorne.

Wheeler's strategy worked splendidly. Within a year of his New York talk, the idiom gradually began to be used in both newspapers and the scientific literature—although for a while at first it was written down as "the black hole," an expression so exotic it needed to be held at a distance within quotation marks. Some, like Richard Feynman, thought the term was obscene. "He accused me of being naughty," Wheeler recalled in his autobiography, *Geons, Black Holes, and Quantum Foam: A Life in Physics.* But Wheeler was attracted to its link to other physics terms, such as "black body," an ideal body that absorbs all the radiation that falls upon it and emits all the energy it absorbs. A black hole does the former but not the latter. It seemingly emits nothing . . . zip . . . nada (more on this in Chapter 30). We look in and see only a dark emptiness. "Thus *black hole* seems the ideal name," concluded Wheeler. Moreover, it fit the very physics of the situation. The collapsed stellar remnant, with its infinite density, was literally digging a hole—a bottomless pit—into the flexible fabric of space-time.

"The advent of the term black hole in 1967 was terminologically trivial but psychologically powerful," said Wheeler.

"After the name was introduced, more and more astronomers and astrophysicists came to appreciate that black holes might not be a figment of the imagination but astronomical objects worth spending time and money to seek." The black hole had finally made it into the big time.

CHAPTER TEN

Like This World of Ours

Confirming that extrasolar planets do exist

IN 2017 an international team of astronomers thrillingly revealed, after examining a collection of data gathered by both NASA's Spitzer Space Telescope and an array of telescopes around the world, that they had found an extrasolar planetary system with at least seven members—all roughly the size of the Earth. These newfound celestial bodies were closely circling a red, Jupiter-sized star known as TRAPPIST-1. The star had been named after the TRAnsiting Planets and PlanetesImals Small Telescope network in Chile and Morocco, which first encountered this extrasolar system. At least three of TRAPPIST-1's rocky planets are likely to harbor liquid water, but so could all seven.

More exciting is that these terrestrial-like worlds are located a relatively scant thirty-nine light-years away in the direction of the Aquarius constellation. In cosmic terms, that's practically next door. Such proximity will allow astronomers to achieve one of their fondest dreams: eventually using current and future telescopes to study the planets' atmospheres

An artist's concept of TRAPPIST-1's seven planets based on data
about their diameter, mass, and distance from the host star.
(NASA/JPL-Caltech)

in search of gases conducive to life, such as oxygen, ozone,
and carbon dioxide.

According to the Extrasolar Planets Encyclopedia, the
number of extrasolar planets so far revealed in our galaxy now
totals in the thousands. The TRAPPIST system was only one
of the latest finds in the burgeoning field of exoplanetary
astronomy.

Although this is a rather new scientific field, speculation
that planetary systems circle other stars started long, long
ago—in ancient times. In the fourth century BCE, the
Greek philosopher Epicurus, in a letter to his student
Herodotus, surmised that there are "infinite worlds both like
and unlike this world of ours." As he believed in an infinite
number of atoms careening through the cosmos, it only
seemed logical that they'd ultimately construct limitless
other worlds.

The noted eighteenth-century astronomer William Herschel, too, conjectured that every star might be accompanied by its own band of planets but figured they could "never be perceived by us on account of the faintness of light." He knew that a planet, visible only by reflected light, would be lost in the glare of its sun when viewed from afar.

But astronomers eventually realized that a planet might be detected by its gravitational pull on a star, causing the star to systematically wobble like an unbalanced tire as it moves through the galaxy. Starting in 1938, Peter van de Kamp at Swarthmore College spent decades regularly photographing Barnard's star, a faint red dwarf star located six light-years away that shifts its position in the celestial sky by the width of the Moon every 180 years, faster than any other star. By the 1960s, van de Kamp got worldwide attention when he announced that he did detect a wobble, which seemed to indicate that at least one planet was tagging along in the star's journey. But by 1973, once Allegheny Observatory astronomer George Gatewood and Heinrich Eichhorn of the University of Florida failed to confirm the Barnard-star finding with their own, more sensitive photographic survey, van de Kamp's celebrated claim of detecting the first extrasolar planet disappeared from the history books.

The wobble technique lived on, however, in another fashion. Astronomers began focusing on how a stellar wobble would affect the star's light. When a star is tugged radially toward the Earth by a planetary companion, the stellar light waves get compressed—that is, made shorter—and thus shifted toward the blue end of the electromagnetic spectrum. When pulled away by a gravitational tug, the waves are extended and shifted the other way, toward the red end of the spectrum. Over time, these periodic changes in the star's light can become discernible, revealing how fast the star is moving back and forth due to planetary tugs.

In 1979, University of British Columbia astronomers Bruce Campbell and Gordon Walker pioneered a way to detect velocity changes as small as a dozen meters a second, sensitive enough for extrasolar planet hunting to begin in earnest. Constantly improving their equipment, planet hunters were even more encouraged in 1983 and 1984 by two momentous events: the Infrared Astronomical Satellite (IRAS) began seeing circumstellar material surrounding several stars in our galaxy; and optical astronomers, taking a special image of the dwarf star Beta Pictoris, revealed an edge-on disk that extends from the star for some 37 billion miles (60 billion kilometers). It was the first striking evidence of planetary systems in the making, suggesting that such systems might be common after all.

The first indication of an actual planet orbiting another star arrived unexpectedly and within an unusual environment. In 1991, radio astronomers Alex Wolszczan and Dale Frail, while searching for millisecond pulsars at the Arecibo Observatory in Puerto Rico, saw systematic variations in the beeping of pulsar B1257+12, which suggested that three bodies were orbiting the neutron star. Rotating extremely fast, millisecond pulsars are spun up by accreting matter from a stellar companion. So, this system, reported Wolszczan and Frail, "probably consists of 'second generation' planets created at or after the end of the pulsar's binary history."

The principal goal for extrasolar planet hunters, though, was finding evidence for "first generation" planets around

stars like our Sun—planets that formed from the stellar nebula itself as a newborn star is created. That long-anticipated
event at last occurred in 1994 when Geneva Observatory astronomers Michel Mayor and Didier Queloz, working from
the Haute-Provence Observatory in southern France, discerned the presence of an object similar to Jupiter orbiting 51
Pegasi, a sunlike star forty-five light-years distant in the constellation Pegasus. They first revealed their discovery at a
conference in Florence, Italy, and their fellow astronomers
declared it a "spectacular detection." Unlike our own solar
system, this extrasolar planet is located a mere four and a half
million miles (seven million kilometers) from its star (far closer than Mercury is to our Sun) and completes one orbit every
four days. Planet hunters had assumed it would take years of
collecting data before detecting the subtle and gradual stellar
wobbles caused by a planet orbiting its parent star, but the
small orbit of 51 Pegasi b enabled them to spot its variations
quickly.

Other discoveries followed swiftly. Geoffrey Marcy and
R. Paul Butler, then both at San Francisco State University
and friendly competitors of the Geneva observers, had been
gathering radial velocity data at Lick Observatory since 1987.
Searching through their records, they found evidence for a
planet similar to 51 Pegasi b, a body at least seven times
the mass of Jupiter closely circling within 40 million miles
(64 million kilometers) of the star 70 Virginis.

These finds challenged theorists, who had not imagined
giant planets with eccentric orbits so close to their sun. These
unusual planets, though, were quickly overshadowed by a
simultaneous discovery by Marcy and Butler—a large planet

orbiting 47 Ursae Majoris at a more distant 200 million miles (around 320 million kilometers). This companion of 47 Ursae Majoris thus gained special distinction for being more "reminiscent of solar system planets." And by 1999, Butler, Marcy, and several colleagues found the first multiple planetary system, a trio of planets circling the star Upsilon Andromedae.

The floodgates were opened, and over the succeeding years thousands of exoplanets were (and continue to be) found. While at first only the biggest exoplanets were revealed (as it was easier to detect them), improved technologies and additional planet-hunting methods enabled the discovery of smaller exoplanets, including Earth-like planets like those in the TRAPPIST system. Space-based missions, such as the Kepler space telescope, were especially productive in spotting these extrasolar planetary systems. "We've gone from the early days

An illustration of the possible surface of TRAPPIST-1f, the fifth of the seven planets orbiting the host star.
(NASA/JPL-Caltech)

of thinking maybe there are five or ten other planets out there, to realizing almost every star next to us might have a planet," says astronomer Jennifer Burt at the Kavli Institute for Astrophysics and Space Research at the Massachusetts Institute of Technology (MIT). Indeed, one team of astronomers in 2012 estimated that there might be one or more planets orbiting each and every Milky Way star. That means at least 200 billion potential homes for ET to call. "We conclude," wrote the astronomers in their *Nature* report, "that stars are orbited by planets as a rule, rather than the exception." Astronomers now know with assurance that the solar system is no longer the sole specimen of its species.

Realm of the Galaxies

By the late 1800s and into the next century, astronomers moved outward. They began to map the topography of our home galaxy, as well as trace how the galaxies themselves are uniquely arranged through the cosmos. And while Edwin Hubble opened our eyes to the existence of other galaxies, it was a woman, her skills at first overlooked, who began to show how those galaxies can evolve over time.

Astronomers also came to learn how starlight could be used to decipher the universe's makeup, what elements resided in both stars and interstellar space. That hydrogen was the overwhelming prime element was discovered by a young woman graduate student in the 1920s. And another woman later provided the best evidence that regular matter—the stuff of stars, planets, and us—is not the major component of the universe. Instead, some unknown "dark matter" is five times more abundant.

Meanwhile, the theories of the great Einstein recast our vision of the universe. We no longer consider the universe as serene, but rather violent, often powered by Einsteinian objects with amazing energies.

All the while, new tools arrived over the past century to better explore the cosmos, instruments taking us beyond the visible-light spectrum. Astronomers now go underground to capture neutrinos emitted from distant events. They gather radio waves, cosmic rays, infrared radiation, X-rays, and gamma rays. They are even detecting the ripples generated in the very fabric of space-time as both black holes and neutron stars collide millions and billions of light-years away.

CHAPTER ELEVEN

Our Spiraling Home

The difficult task of mapping the Milky Way

IT'S an odd quirk of astronomy. Telescopes can peer billions of light-years outward, allowing us to observe a plethora of galaxies and to map their lacelike distribution through space and time with exquisite precision. Distant quasars have been thoroughly examined—by gathering their emissions from radio to gamma rays. And the European Space Agency's Planck satellite provided one of the best baby pictures of the cosmos yet, an image depicting the universe when it was a mere 400,000 years old. It appears that the definitive cosmic atlas is within the grasp of astronomers.

And yet the topography of our *local* celestial landscape (that is, within only tens of thousands of light-years) remains frustratingly murky. What seems as if it should be the easiest structure to trace—that of the Milky Way, our home galaxy— is just the opposite. It's like owning the best globe of the world, with your hometown missing. There's a simple reason for this conundrum: our solar system is embedded inside the dusty plane of the Milky Way. Such a position makes viewing our galaxy's exact configuration a difficult task. Try discerning the pattern on a piece of china with your eyes level to the edge of the plate. That's what astronomers have long confronted when trying to map the Milky Way. How can you peer through that disk, full of dust and gas?

To find an answer, astronomers started with a hunch—a reasonable one at that. Since the disks of other galaxies displayed a beautiful spiraling architecture, they assumed that the Milky Way, too, has massive arms that wrap themselves around the galactic hub like coiled streamers.

By the 1930s, identifying the Milky Way's spiral arms became a top item on astronomers' agenda. At first they tried just counting stars, all the ones in sight, hoping denser concentrations in the tally would outline the arms. But, alas, they experienced little success. It took World War II, oddly enough, for astronomers to come across a new approach for solving this problem.

Because of the fear that the Japanese might attack the west coast of the United States, the Los Angeles area was blacked out nightly during the conflict. This war-imposed veil of darkness was heaven for one particular astronomer working at the nearby Mount Wilson Observatory, which operated the biggest telescope in its day: the Hooker telescope, with its 100-inch-wide (2.5-meter-wide) mirror. While many observatory staffers had temporarily left to carry out war work, German-born Walter Baade was designated an "enemy alien" and restricted to the Pasadena area. That meant he had almost unlimited time on the 100-inch, allowing him to get the best look ever at the Andromeda galaxy, the spiral galaxy closest to us, at a distance of 2.5 million light-years.

Pushing the telescope to its limits over months of observations, Baade came to recognize that highly luminous blue and blue-white supergiant stars, along with bright gaseous nebulae, tended to reside *only* in Andromeda's spiral arms, acting much like the lights lining an airport runway. The reason spiral arms stand out is because they are regions where young, hot stars are forming.

The stars making up an arm are not permanently connected, as if part of a ropelike structure attached to a galaxy's center. Rather, that appearance reflects underlying density or shock waves that travel through the galaxy's disk of gas and foment star formation. As the disk's gas passes through this compression wave during its rotation around the galactic center, the material gets squeezed, huge clouds form, and within several million years big new stars turn on to illuminate the density wave's spiraling structure. It's like a cosmic traffic jam. The highly luminous stars are so short-lived, however, that they die off by the time they move out of the traffic tie-up. This mechanism behind a galaxy's spiraling structure wasn't identified until the 1960s, but nonetheless Baade had still found the perfect objects to delineate a spiral galaxy's arms.

Soon after the war, others began applying this newfound knowledge to our own galaxy. Astronomer William W. Morgan at the Yerkes Observatory in Wisconsin got a head start on the problem, as he had already been carrying out a spectral study of the Milky Way's brilliant supergiant stars. He first teamed up with Jason Nassau at the Warner and Swasey Observatory in Ohio, and together they pinpointed the positions of some 900 supergiants. Less than 6 percent of these stars had their distances reliably nailed down, yet this evidence, scanty as it was, suggested that a spiral arm might be running from the constellation Carina over to Cygnus in our local solar neighborhood. It was a start.

Soon after, Morgan joined forces with two student assistants, Stewart Sharpless and Donald Osterbrock, to push the survey even further. Along with tracking down blue and blue-white stars, they also plotted the distribution of luminous nebulae (notable for their energized hydrogen). To quickly spot the nebulae, the two students set up a special camera that was originally designed as a wide-angle projector for training aerial gunners during World War II. The dozens of photographic plates they took, revealing many new nebulae for Morgan to analyze, provided the breakthrough.

With the additional data, segments of two spiral arms could be reliably traced. One arm (labeled Orion) passed within 1,000 light-years of the Sun; the other (Perseus), located farther from the galactic center, was at its closest point to us some 6,500 light-years away. There was also the hint that a third spiral arm (Sagittarius) swept closer to the center of the Milky Way, about 5,000 light-years away from us.

The Yerkes team announced its findings at a 1951 meeting of the American Astronomical Society held in Cleveland, Ohio. Morgan presented a handmade model of the spiral arms, which used cotton balls to depict the positions of the bright nebulae. This map was far from complete, because it's difficult for an optical telescope to peer much farther into the dust- and gas-filled plane of the Milky Way. But that didn't dampen the reception Morgan's work received at the astronomical conference.

"Astronomers are usually of a quiet and introspective disposition," University of California astronomer Otto Struve later wrote. "They are not given to displays of emotion. . . . But in Cleveland, Morgan's paper on galactic structure was greeted by an ovation such as I have never before witnessed. Clearly, he had in the course of a 15-minute paper presented so convincing an array of arguments that the audience for once threw caution to the wind and gave Morgan the recognition which he so fully deserved."

There was clapping of hands and stomping of feet. And why not? The Yerkes astronomers were providing the first map (partial as it was) of our cosmic "hometown." A problem that astronomers had struggled with for decades, astronomy historian Owen Gingerich has written, had "finally found its solution by a quite different avenue from the numerical star-counting procedures."

And when it rains, it pours. Within two years, the spiraling segments were confirmed and extended with the use of a new instrument available to astronomers—the radio telescope, which could penetrate farther through the Milky Way's dust and haze by tuning in to a radio frequency emitted by hydrogen gas.

The map is still incomplete, but some overall patterns are emerging. For one, there is now strong evidence that the Milky Way galaxy is a barred spiral, which means its center is extended like a bar rather than bulbous (as in the Andromeda galaxy). More than two-thirds of present-day spirals have a center bar, a structure that likely evolves as a spiral galaxy matures.

And from the 1950s into the 1990s, continuing surveys revealed further sections of our galaxy's spiraling arms, piece by piece. By connecting the dots, astronomers came to believe that there were four gently curving arms, neatly arranged around the Milky Way's center.

Sun

An illustration of the most up-to-date information on the Milky
Way's structure as a barred spiral. The Sun resides in the Orion
spur, about two-thirds of the way from the galactic center.
(NASA/JPL-Caltech/R. Hurt [SSC/Caltech])

Infrared images taken by NASA's Spitzer Space Telescope, released in 2008, changed that assumption, however. It
now looks like our galaxy has two dominant arms. Each orig-
inates from an opposite end of the central bar and then bends
outward, swirling nearly completely around our galaxy's core.
One of them, the Perseus arm, was partially seen by Morgan
in 1951. The other is known as the Scutum-Centaurus arm.

What happened to Morgan's other spiral-arm sightings?
Much like the earliest maps of the New World, the cartography

has been altered with better resolution. Morgan's Sagittarius arm has been demoted to a more minor appendage, while the Orion arm (where the Sun resides) is now known to be a mere "spur." Our view of the Milky Way arises within a smaller concentration of stars and nebulae that is positioned between the two major arms. We sit amid glory—but at a smaller table.

CHAPTER TWELVE

The Woman Who Chased Galaxies

The historic legacy of a "mere Dallas housewife"

B Y the 1930s Edwin Hubble was well into his search for far-off galaxies. Using the great 100-inch telescope atop California's Mount Wilson, he could see out to distances of a few hundreds of millions of light-years. But beyond that, the smudges on his photographic plates were dim, fuzzy, and next to impossible to identify. "There," wrote Hubble in 1936, in his classic book *The Realm of the Nebulae*, "we measure shadows, and we search among ghostly errors of measurement for landmarks that are scarcely substantial." Ever since, astronomers have struggled to trace the evolution of galaxies back through space-time—not just hundreds of millions of light-years outward, but billions.

Hubble himself saw no changes over the relatively shallow span he surveyed. Galaxies "are enormous systems, and it is reasonable to suppose that their evolution is correspondingly slow," he concluded. And this became the prevailing view for the next three decades. Astronomers just assumed that all the galaxies—every spiraling pinwheel and bulbous elliptical—formed fairly quickly after the Big Bang and then coursed serenely through the cosmos, changing very little over the eons. They had no reason to doubt this. Given how far back astronomers could see at the time (which wasn't very far), distant galaxies looked pretty much like the galaxies right by us.

Cosmologists in those early days depended on this axiom of constancy. Their prime motivation for tracking galaxies at all was their insatiable desire to learn the universe's fate. They were little interested in the galaxies themselves; galaxies were simply markers, convenient spots in space to discern the rate of the universe's expansion. By comparing the speeds of galaxies in earlier epochs with those of today, they hoped to judge whether galaxies were slowing down enough to someday stop in their tracks by the pull of gravitation and eventually fall back toward one another in a "Big Crunch." On the other hand, maybe they were flying outward at an unstoppable speed, keeping the cosmos forever open.

Using galaxies for this cosmological measurement was a fine idea, as long as galaxies could be thought of as immutable objects that drifted on in tranquil isolation. By maintaining a uniform size and brightness over time, the galaxy could be used as a yardstick. An astronomer estimated a galaxy's distance by measuring its luminosity and angular width on the sky. As observers peered deeper and deeper into space, viewing the universe as it was in the past, they assumed that ever more distant galaxies would appear dimmer and smaller in a systematic fashion.

But what if a galaxy gets either brighter or fainter with age? What if it changes its shape from eon to eon? Then all bets are

Beatrice Tinsley.
(Astronomical Society of the Pacific, courtesy AIP Emilio Segrè
Visual Archives, *Physics Today* Collection)

off, and the universe's destiny is far harder to determine in this way. Cosmologists learned this unhappy fact—galaxies change!—by the 1970s, and the person primarily responsible for establishing this new principle was Beatrice Tinsley.

Born in England in 1941 and raised in New Zealand, where her family moved after World War II, Tinsley did a master's thesis in solid state physics. Soon after, in 1963, she moved to the United States when her husband, physicist Brian Tinsley, garnered a research job in Dallas. By then she had plans to pursue a doctorate, but now she wanted to specialize in her long-standing passion—cosmology, a choice that hadn't been available to her in New Zealand.

Though some judged Tinsley as a mere Dallas housewife with no experience in astronomy, her top-notch academic record convinced the head of the astronomy department at the University of Texas, Austin, to take a chance on admitting her, even with the added burden of her commuting the two hundred miles (320 kilometers) from Dallas to Austin.

Initially Tinsley planned to take part in the long-standing cosmological pursuit of deciding whether the universe was open or closed. But as she examined all the observables in this line of work—the diameters of clusters of galaxies, galaxy magnitudes, galaxy sizes—one question kept diverting her: How were the galaxies changing over time? How were they evolving? That information was crucial to finding an answer to the universe's fate.

At that point she chose the problem that became her dissertation: actually simulating the evolution of a galaxy. Setting up a numerical model, she would track its changes in color and brightness over billions of years as the stars within it are born, fiercely radiate, and then inevitably die. It was an ambitious task, as numerical simulations were grueling in this primordial era of computing.

No one before had ever tackled such a problem in great detail. It has been described as "one of the boldest graduate thesis projects ever undertaken." Tinsley had to set up an initial population of stars and then decide how quickly they would die and how soon new stars would be generated to take their place. And no one yet knew for sure whether a galaxy's brightness depended more on the collective light emanating from its numerous long-lived, low-mass stars or from its scarcer—but far brighter—short-lived, massive ones. Tinsley constructed her model based on the best theoretical and observational evidence available at the time.

After her dissertation was completed in 1967, she continued to refine her models over the years, each simulation concluding that galaxies can undergo substantial evolution through time, far more than astronomers had previously thought. A young galaxy starts out bright and blue, when its resources of gas to form stars are at their peak, and then gently reddens with age and dims considerably as the stars age and die over the eons.

Some more senior authorities at first took issue with these conclusions, but eventually her findings encouraged observers to start pushing outward with their telescopes to discern her predicted galactic evolution firsthand. As a consequence, Tinsley's papers began to be cited in dozens of scientific publications. Yale University took notice in 1975 by offering her a professorship, a post she had been unable to secure years earlier (to her great frustration) in either Texas or elsewhere. The woman once regarded in Dallas as "Brian Tinsley's clever wife, rather than as a scientist in her own right," according to science historian Joann Eisberg, had proved that people had vastly underestimated her talent.

It didn't take long for astronomers to get direct confirmation of Tinsley's theoretical findings. In 1977 astronomers Augustus Oemler Jr. and Harvey Butcher used the 84-inch (2.1-meter) telescope on Kitt Peak in southern Arizona to analyze the light emanating from two galaxy clusters, now known to be situated some five billion light-years away (hence five billion years back in time). What they saw matched Tinsley's prediction: the galaxies in both clusters were radiating

A Hubble Space Telescope photo of galaxy cluster
CL0024+1654, earlier studied by Augustus Oemler
and Harvey Butcher in 1977 to prove galaxy evolution.
(NASA/ESA/H. Lee & H. Ford [Johns Hopkins])

more blue light than the more reddish clusters near us today,
likely because there were more blue, energetic galaxies in
those clusters than the clusters near us today. No longer mere
markers, distant galaxies were now viewed as fascinating and
evolving cosmic creatures worthy of study all on their own.

All those efforts ushered in a new era in extragalactic research. Gradually Hubble's "shadows" began to disappear
as new and improved instrumentation allowed the early universe to come into better focus. Faraway galaxies that had
been smudges in Hubble's day are being viewed today with

impressive clarity. And what astronomers are seeing is that galaxies over time can exhibit diverse personalities. Some do move serenely through the cosmos, evolving internally as Tinsley calculated, but astronomers now know that many can also change more recklessly. Galaxies may collide, merge, sideswipe one another, or gobble up unwitting passersby. The resultant galaxy-wide temblors often trigger the birth of millions of stars. It is a wondrously invigorating picture of extragalactic affairs, in which galaxies evolve, either dimming or brightening as they age, owing to *outside* influences.

Sadly, Beatrice Tinsley witnessed very little of the new era she inspired. In 1978 a lesion on her leg was diagnosed as melanoma, a malignant skin cancer. While continuing to teach and carry out research, she underwent extensive radiation and chemotherapy, but ultimately the treatments were unsuccessful. She died in March 1981 at the age of forty. Writing in *Physics Today* a few months later, Sandra Faber of the University of California, Santa Cruz, observed that Tinsley had "changed the course of cosmological studies."

Two weeks before her death, while hospitalized in the Yale infirmary, Tinsley submitted her last scientific paper to the *Astrophysical Journal*. No longer able to use her right hand, she had written it with her left. The article advanced her work on galaxy evolution and was published the following November without revision by the editors.

CHAPTER THIRTEEN

Stuff of the Heavens

How light taught astronomers what the universe is made of

I N 1999 NASA launched a spacecraft called *Stardust* into the heavens to capture just what its name suggested: matter from outer space that likely originated from long-dead stars, whose remnants provided the material out of which our solar system formed.

In its years-long journey, eventually covering billions of miles as it orbited the Sun, *Stardust* flew through a stream of interstellar dust, as well as the coma of Comet Wild 2, collecting specks of matter onto its tennis-racket-wide aerogel collector. In 2006 the probe returned to Earth's vicinity and ejected its precious cargo. Safely nestled in a special capsule, the payload landed in Utah's Great Salt Lake Desert in the dead of night. Transported to NASA's Johnson Space Center in Houston, Texas, this cosmic treasure—tens of thousands of microscopic and submicroscopic grains—has been under close scrutiny ever since.

One of the most startling revelations of the dust's analysis was the discovery of glycine, the smallest of the twenty amino acids that serve as vital building blocks for our body's proteins. "The significance of this discovery is that comets must have delivered at least one amino acid to our planet before it had life," said *Stardust* principal investigator Don Brownlee. Other researchers have found nucleic acids, components of DNA and RNA, in meteorites. It's further confirmation that "we are made of starstuff," as Carl Sagan so famously described it in his book *Cosmos*.

That we have such an intimate connection to the cosmos is actually a relatively new revelation. For most of history, astronomers could not be sure that the stuff of the heavens was anything at all like the stuff on Earth. And since outer space was so inaccessible, they figured an answer would be forever out of their reach. The French philosopher Auguste Comte was so confident in this judgment that in 1835 he boldly asserted that "we would never know how to study by any means [the stars' and planets'] chemical composition, or their mineralogical structure." That declaration is one of the most infamous misstatements in the history of science. What Comte did not anticipate was the development of new techniques that—in less than three decades—would sweep away his ill-timed conclusion.

The turnabout primarily happened when Gustav Kirchhoff, a professor of physics at the University of Heidelberg, and chemist Robert Bunsen, creator of the famous laboratory burner, teamed up in 1859 and demonstrated how to identify substances by the specific colors of light they emit during chemical reactions or when burning. Whenever energized and viewed through a spectroscope, each element could be recognized by a unique set of colored lines it displayed.

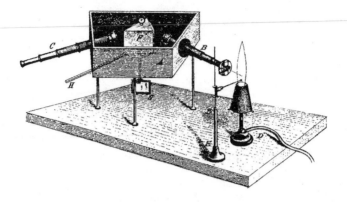

The spectrometer used by Gustav Kirchhoff
and Robert Bunsen.
(From *Annalen der Physik und der Chemie*, 1860)

Soon the two collaborators realized that such spectral
fingerprints could be effectively studied whether the light
originated from a distance of one foot within a laboratory or
from millions of miles away. That insight may have been
prompted by a fire that erupted in the nearby city of Mannheim
and was visible across the Rhine river plain from their laboratory window. Upon directing their spectroscope at the flames,
Kirchhoff and Bunsen discerned the strong green emission
of barium in the roaring blaze, as well as the distinctive red
signature of strontium. Sometime later, while they were
strolling together through the wooded hills near Heidelberg,
Bunsen wondered if they could analyze the Sun's light in a
comparable fashion. "But people would say we must have
gone mad to dream of such a thing," he declared.

Kirchhoff, though, had no such qualms. By 1861 he
had turned his spectroscope to the heavens and identified a

number of elements in the Sun's atmosphere, including sodium, magnesium, calcium, chromium, iron, nickel, copper, zinc, and barium. Within a few years, other astronomers, such as Angelo Secchi in Italy and William Huggins in England, reported finding similar elements in such distant stars as Aldebaran, Betelgeuse, and Sirius. Here was definitive proof that the chemical elements of the Earth were indeed identical to those of the cosmos. The long-standing Aristotelian belief that celestial matter was somehow different from the terrestrial elements was abolished once and for all.

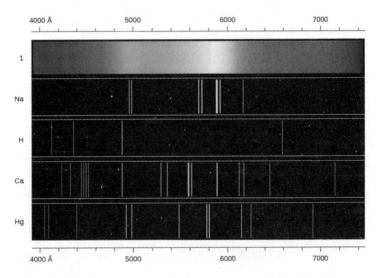

At the top, a continuous spectrum that runs from violet (left) to red (right). Below that, the specific spectral "fingerprints" of sodium (Na), hydrogen (H), calcium (Ca), and mercury (Hg).
(OpenStax, Chemistry. OpenStax CNX, https://opentextbc.ca/chemistry/. June 20, 2016. Copyright 2016 by Rice University. License at https://creativecommons.org/licenses/by/4.0)

Huggins, for one, was elated by these discoveries and couldn't help but speculate on what this implied. In 1864 he and his collaborator, W. Allen Miller, wrote, "It is remarkable that the elements most widely diffused through the host of stars are some of those most closely connected with the constitution of the living organisms of our globe . . . that at least the brighter stars are, like our sun, upholding and energizing centres of systems of worlds adapted to be the abode of living beings."

It wasn't the first time that scholars speculated about life on extrasolar planets, but the new astrochemical data now made it more than a theoretical fantasy.

A century later, some researchers became even more ambitious. In 1955, physicist Charles H. Townes, who would later win a Nobel Prize for the invention of the maser (the microwave precursor to the laser, which emits electromagnetic radiation at higher-frequency, visible wavelengths), was invited to address an international symposium on radio astronomy in England. His topic: the possibility of detecting celestial substances, other than simple elements, via their radio emissions. Townes, a renowned molecular spectroscopist, suggested that elements were likely linking up and forming actual molecules out in space—molecules that emitted intense radio waves. Among the candidates he named were carbon monoxide (CO, the dangerous stuff of car exhaust), ammonia (NH_3), water (H_2O), and the hydroxyl radical (OH, the oxygen-hydrogen combination that distinguishes all alcohols and is important in atmospheric chemistry).

The response to Townes's talk was tepid, however. Most astronomers at the time were convinced that such molecules were too rare to seek out. Optical astronomers had already recognized a few molecular species in space, such as the methylidyne and cyanide radicals (CH and CN), but theorists were sure that, once formed, such molecules quickly got destroyed by ultraviolet and cosmic rays. Why devote precious radio telescope time to tracking scarce specimens, which everyone assumed were unimportant to astronomical processes? One of Townes's colleagues cautioned him that such a search would be "hopeless."

Fortunately, a few MIT radio astronomers didn't heed those warnings and looked anyway. In 1963 they found hydroxyl radicals screaming out at a frequency of 1,667 megahertz in the supernova remnant Cassiopeia A. Five years later, Townes himself, along with coworkers at the University of California at Berkeley, recorded the radio cries of both ammonia and water molecules in the galactic center.

A race quickly ensued to snare the next new cosmic molecules. By 1973 nearly thirty were identified; the total today is more than 150—from acetone and hydrogen cyanide to formaldehyde, methane, and nitrous oxide (laughing gas). Astronomers handed out cases of liquor to settle bets once ethyl alcohol was detected in 1974. It's been estimated that 10^{22} (that's one followed by twenty-two zeros) fifths, at 200 proof, reside in the gas cloud where the alcohol was first detected. Of course, the molecules are spread out so thinly in space that you'd have to distill a volume as big as the planet Jupiter to get one stiff drink.

These assorted molecules barely register as pollutants in our galaxy. Only one molecule of ammonia, for example, forms for every 30 million molecules of hydrogen. Yet scarce as these molecules are, their strong radio signals allow astronomers to better map both the Milky Way and the universe.

Hydrogen peroxide, the hair-bleaching agent, was uncovered just several years ago (who knew the cosmos secretly desired to be a blonde?). Using a submillimeter-radio-wave telescope perched on a high desert plateau in the Chilean Andes, an international team of astronomers found traces of the chemical in a dense cloud of gas and dust near the star Rho Ophiuchi, some four hundred light-years distant. Hydrogen peroxide is formed when two hydrogen atoms link up with two oxygen atoms (H_2O_2). Both elements are critical for life as we know it. Moreover, take just one oxygen atom out of hydrogen peroxide and you get water (H_2O). So, further study of hydrogen peroxide's chemistry out in deep space may help astronomers better understand the formation of water in the universe.

Molecule by molecule—from water to glycine—astronomers are proving that the foundations for life on Earth may have been put into place before our planet even formed nearly five billion years ago.

CHAPTER FOURTEEN

Recipe for the Stars

When a graduate student discovered abundant
hydrogen in stellar spectra, she was bullied
into suppressing her results

THE resolve to pursue science was never an easy choice for a young girl in the Edwardian age. Yet one could, by dint of talent, drive, and the careful choice of one's parents and social class, overcome the more blatant barriers to a scientific education. Cecilia Helena Payne, in her later years as an astronomer at the Harvard College Observatory, could point to her mild childhood confrontation with the female stereotype gently enforced by the administration of the church school she attended in London. The female principal told her that she would be prostituting her gifts by embarking on a scientific career. But Payne, born in 1900 in Wendover, England, was descended from a family of scholars and historians, and she eagerly unearthed books on botany, chemistry, and physics in the extensive library at her family home. Her father, a barrister, died when she was four, but her mother, an accomplished musician, carefully guided Payne's education. A simple move to a new and more modern school enabled her to immerse herself in scientific studies.

Payne flourished at the new school and became enchanted by the prospect of life as a scientist. "I knew, as I had always known," she confessed much later in her autobiography, *The Dyer's Hand*, "that I wanted to be a scientist [but] was seized with panic at the thought that everything might be found out before I was old enough to begin!" Of a room set aside for science instruction, she once recalled: "The chemicals were ranged in bottles round the walls. I used to steal up there by myself . . . and sit conducting a little worship service of my own, adoring the chemical elements. Here were the warp and woof of the world."

Without much ado Payne stayed the scientific course in high school, and in the autumn of 1919, shortly after World War I ended, she entered Newnham College at the University of Cambridge.

Payne's arrival at Cambridge as an undergraduate coincided with a tremendous upheaval in the understanding of the physical world, when the physics community was reeling from the startling new discoveries thrust upon it. Until the end of the nineteenth century, scholars generally had thought of the universe as a smooth-running clock, and the science of the day was essentially guided by the same principle. The success of Newton's equations of motion had led to a smug assurance that every phenomenon in the cosmos could ultimately be explained mechanically. But nature was not following that script, and things quickly went awry when theorists tried to apply the mechanistic laws of classical physics to the workings of the atom.

For several decades astronomers had been identifying elements in the heavens by comparing their spectral emissions and absorptions with those of glowing gases in the laboratory. The mechanism that gave rise to the light, however, was a complete mystery. Then, in 1913, the Danish physicist Niels Bohr deduced that an atomic spectrum is generated as the electrons in an atom jump from one orbit to another, emitting or absorbing bursts of light along the way. That theory enabled Bohr to calculate the specific colors of light that should be absorbed or emitted by hydrogen, corresponding to the difference in energy between a high electron orbit and a lower one in that atom. Bohr's predictions matched the observed spectrum of hydrogen almost perfectly. On hearing the news, Einstein is said to have remarked, "Then this is one of the greatest discoveries ever made."

Payne had the wit and tenacity to become one of the first astronomers to apply the new laws of atomic physics to astronomical bodies. In the course of her painstaking thesis calculations, which drew heavily on the new physics, she uncovered the first hint that hydrogen, the simplest element, is the most abundant substance in the universe. The reverberations of that plain fact still echo in astronomy. Here is the fuel for a star's persistent burning; here is the gaseous tracer that enables radio astronomers to probe a dark, long-hidden universe; here is the remnant debris from the first few minutes of creation. Payne's discovery did no less than change the face of the material cosmos.

And yet Payne's name (and equally, her married name, Payne-Gaposchkin) is missing from most astronomy books. One can debate the point—for the evidence is not unambiguous—but her failure to gain the very first rank among astronomers seems to have been caused by the forces of sexual inequality. At the last minute, pressured by her more conservative superiors, she virtually retracted her discovery of stellar hydrogen and published a statement far less definitive than what she actually believed. Her findings were so radical, so different, that she was pushed into softening her thesis. Ironically, the professor who most influenced her to back down eventually confirmed her original suspicions and published the seminal paper on the hydrogen makeup of the stars. Payne has been described as the most eminent woman astronomer of all time. Her doctoral degree was the first ever granted to a student at the Harvard College Observatory (the university's physics department had refused to accept a woman candidate). But her failure to achieve recognition for one of the most important advances in astrophysics tells much about the pressures on women scientists as they make their way in a man's world.

At first in Cambridge Payne leaned toward a career in botany, a childhood passion. But she made sure to add physics and chemistry to her studies. She found the renowned physicist Ernest Rutherford, who was then conducting some of his most creative experiments at the Cavendish Laboratory, "irresistible." "He was always on the horizon, Payne recalled, "a towering blond giant with a booming voice."

The pivotal decision to dedicate her life's work to astronomy came one winter night in 1919. Four years earlier Einstein had introduced his general theory of relativity, which, among other things, predicted that beams of starlight grazing the Sun would get bent by a slight but detectable amount.

Harvard astronomer Cecilia Payne-Gaposchkin.
(Smithsonian Institution Archives. Image #SIA2009-1325)

The bending had been predicted in earlier theories, but in general relativity it was calculated to be twice as large because of the curvature of space-time in the vicinity of the Sun. With safe travel restored after World War I, British astronomers eagerly mounted two expeditions to test Einstein's conjecture. On May 29, 1919, from sites in northern Brazil and on the small island of Principe, off the coast of western Africa, the investigators photographed stars near the edge of the Sun during a total solar eclipse.

Arthur Eddington, then the foremost astronomer at Cambridge, was a member of the Principe brigade, and he presented the results of the fabled undertaking in the Great

Hall of Cambridge's Trinity College. The event was sold out, but Payne had miraculously chanced upon a ticket. There Eddington reported that the gravitational deflection of the stellar rays agreed closely with Einstein's calculations.

For Payne it amounted to a religious conversion. She deserted the life sciences and informed the school authorities that she would be devoting her studies to the physical sciences. She was already aware of the problems in her chosen disciplines. Shy and awkward, Payne trembled when she had to sit alone in the front row at Rutherford's lectures, the required seating arrangement for any lone woman in a sea of male students. Her physics lab instructor would often shout at the female students: "Go and take off your corsets!" certain as he was that the steel frameworks of the corsets would disturb his magnetic equipment. But one night, when the Cambridge Observatory was open to the public, Payne encountered Eddington personally. She blurted out, "I should like to be an astronomer." "I can see no insuperable objection," he replied, and he proceeded to widen her opportunities for research.

Payne, faced with the prospect that her only job in England after she completed her degree would be teaching science at a girls' school, was advised to go to the United States. There women had better opportunities in astronomy than they did in England. Another spellbinding lecture, this one given in London by a young sandy-haired Harvard astronomer named Harlow Shapley, prompted Payne to set her sights on Massachusetts, home to the largest storehouse of astronomical data in the world. Her Cambridge professors and colleagues were highly supportive in their recommendations. Leslie J. Comrie, who was by then teaching at Swarthmore

College in Pennsylvania, wrote to Shapley that Payne was "the type of person who, given the opportunity, would devote her whole life to astronomy" and that "she would not want to run away after a few years' training to get married." Payne got a fellowship.

Shapley had hoped Payne would work at Harvard on determining stellar brightnesses, a fairly routine endeavor. But Payne was more intrigued by the physical interpretation of stellar spectra, a more theoretical pursuit and the Harvard specialty. Swift to grasp and apply new ideas, she knew that the work of the young Indian physicist Meghnad Saha could serve as a powerful diagnostic of a star's surface conditions. Saha had recognized that each element stands out vividly in a stellar spectrum only at a particular temperature and pressure, usually when the conditions are intense enough to ionize those elements, stripping the atoms of some of their outer electrons. Otherwise, the element would remain essentially hidden from view. Saha's ionization theory gave the first physical explanation for the striking distinctions among the various observed kinds of stellar spectra. In many ways, Saha's realization marks the beginning of modern astrophysics.

Payne was excited by the prospect of verifying Saha's theory with the myriad spectra available in the Harvard astronomical plate collection. She compared her research to an archaeological dig; the data were "bones to be assembled and clothed with the flesh that would present the stars as complete individuals." Her early training in the systematic classification of plants served her well. She looked at hundreds of spectra (her "celestial flora") and selected certain known spectral lines for inspection. She set up a crude system for estimating the intensities of the spectral features, an arduous task. "There followed months, almost a year I remember, of utter bewilderment," she said. But "nothing seemed impossible in those early days. . . . We were going to understand everything tomorrow."

Gradually, answers did arrive. After days and months of grappling with her treasured plates, the intensities of the lines of silicon in four successive stages of ionization began to make more sense. With that key she determined the temperatures of the hottest stars, and from that day forward silicon was Payne's favorite element.

With the job complete, Payne proceeded to the calculations for which her thesis is most famous: the relative abundances of eighteen elements commonly observed in the atmospheres of various classes of stars. Her guides were Saha's equations and statistical mechanics, especially the seminal work of the English theoretical physicists Edward Milne and Ralph Fowler. She was able to estimate the number of atoms needed to generate a particular spectral feature. Payne was immediately struck that the common elements in the Earth's crust were also present in the stars. For elements such as silicon and carbon in the stars, she even found the same relative proportions as exist on Earth.

Those findings seemed in accord with trends elsewhere in astrophysics. In 1914, some ten years before Payne's work, Henry Norris Russell, then director of the Princeton University Observatory, had compared the most common materials in the Earth's crust with the substances commonly observed in the Sun. To a large degree the solar and terrestrial compositions matched. In the 1890s the American physicist Henry A. Rowland, who had prepared an exquisite map of the solar spectrum, had remarked that if the Earth's crust were heated to searing solar temperatures, its spectrum would probably look much like the Sun's.

But there the similarities ended between Payne's stellar abundances and the relative abundances of the elements on the Earth. Two elements stood out as startling exceptions to the rule: "Hydrogen and helium are manifestly very abundant in stellar atmospheres," Payne reported. Indeed, her results suggested that hydrogen alone could be as many as a million times more plentiful in the stars than it is on the Earth. Helium in the stars, she noted, was about a thousand times more abundant than the heavier elements.

The winds in physics, though, were blowing against Payne's findings. Eddington, the world's expert on stellar structure, had figured that the average atomic weight of a star's material was far greater than that of hydrogen, the lightest gas of all. He was so sure his stellar models would not work with high hydrogen abundances that when he applied the quantum rule known as Kramer's law to the interior of the Sun and came up with extremely high hydrogen abundances, he assumed the law was wrong. No one—except perhaps a young female graduate student—was quite ready to challenge Eddington's theoretical prowess on stellar interiors.

At each major step in her analyses Payne wrote a paper describing her findings, completing half a dozen articles before receiving her doctorate in 1925. But she turned conservative when publishing the results on stellar abundances. "In the stellar atmosphere and the meteorite the agreement is good for all atoms that are common to the two," she wrote. "The outstanding discrepancies between the astrophysical and terrestrial abundances are displayed for hydrogen and helium."

Then, on the verge of recognizing that the two elements make up the bulk of stellar material, and hence the preponderance of the matter in the universe, Payne pulled back.

Why did she hesitate? "She was bullied," contended Jesse Greenstein decades later. Greenstein was a veteran astronomer at the California Institute of Technology, who had once been an old friend and Harvard colleague of Payne's. "All papers at Harvard, unfortunately, had to be approved by the director, Harlow Shapley." In December 1924 Shapley sent Payne's manuscript to Russell, Shapley's mentor and former teacher at Princeton. A whiz at mathematical computations and a lifelong workaholic prone to nervous breakdowns, Russell was in the vanguard of incorporating modern physics into astronomy. Payne respected and feared Russell, who always seemed to speak with the voice of authority.

Russell at first concluded that Payne's findings were "a very good thing." But five weeks later he had second thoughts, and he wrote the young graduate student, "there remains one very much more serious discrepancy. . . . It is clearly impossible that hydrogen should be a million times more abundant than the metals." In an article sent to *The Proceedings of the National Academy of Sciences* in February 1925, Payne withheld her original conclusion and instead wrote that the abundance calculated for both hydrogen and helium "is improbably high, and is almost certainly not real," a statement reiterated in her doctoral thesis. She toed the party line that a star's makeup basically resembles the composition of the Earth's crust.

"Cecilia was a tough cookie," said Greenstein, yet she still acquiesced to Russell's counsel. It is hard for any graduate student to challenge the leaders in the student's field, especially giants such as Russell and Shapley. Without the approval of these two notable astronomers, Payne's thesis would not have been published. Her career and her place in the astronomy community depended on them—and she respected both of them enormously. In fact, Payne had an innocent crush on Shapley, and the two engaged in many long, stimulating scientific discussions, though the director always kept his distance personally. Payne, on the other hand, admitted to a slavish, platonic devotion. "In those days I worshiped Dr. Shapley; I would gladly have died for him," she confessed.

Her autobiography does not elaborate at all on the Russell episode and the controversy over her findings. There is only a vague remark that Russell, whose word could make or break a young scientist, vetoed some of her cherished ideas. Yet although she backed down in print, she held to her conviction. Payne visited Cambridge University shortly after her thesis was completed and informed Eddington in a burst of youthful zest that she believed there was far more hydrogen in the stars than any other atom. "You don't mean in the stars," replied Eddington, "you mean on the stars."

It may be that Russell, Eddington, and Shapley were not being obstinate, just cautious. Atomic physics was exploding just as Payne was writing her thesis, and several solar features, such as the Sun's opacity, were just beginning to be understood. Knowledge of the atom's structure had only recently moved from the visions of Democritus and Dalton to the ones of Rutherford and Bohr. Finally, Payne was working with crude data, and Russell warned her that hydrogen, because of its simplicity (one proton and one electron), might be giving skewed results.

Yet Payne's hesitation at officially naming hydrogen the prime element hardly diminishes her other accomplishments. Her thesis was the first to combine atomic theory, Saha's new equations, and astronomical observations to obtain good estimates of the elemental abundances in the stars, as well as detailed analyses of stellar temperatures and pressures. A number of astronomers would later describe her work as the most brilliant Ph.D. thesis ever written in astronomy. In 1926, at the age of twenty-six, she became the youngest astronomer listed as distinguished in *American Men of Science*. Edwin Powell Hubble, whose observations would soon confirm that the universe was steadily expanding, joked that she was "the best man at Harvard."

But the many recognitions were shallow triumphs. Mainly because of her sex, a professional position worthy of her expertise eluded her. The historian Peggy Aldrich Kidwell of the National Museum of American History in Washington, D.C., who has written extensively on Payne's work and life, points out that women were either ineligible or simply unwanted for posts at colleges with the best observatories. Payne was paid for a time as Shapley's technical assistant while she lectured and conducted research, but she received no official Harvard appointment until 1938. The courses she taught there were not listed in the university catalogue until 1945. In 1956 she was made a full professor—the first woman at Harvard to attain that rank—perhaps twenty years after a man of her achievements would have earned the same position.

Greenstein first met Payne when he was a young Harvard student and she had completed her thesis just a couple of years before. "The obvious discrimination against her as a woman scientist worthy of normal academic recognition exacerbated the stressful life she led," he says. "She was unhappy, emotional. . . . But with me, she was charming and humorous as we exchanged quotations from T. S. Eliot, Shakespeare, the Bible, Gilbert and Sullivan and Wordsworth." She was also, in the words of her daughter, Katherine Haramundanis, a "world traveler, . . . an inspired seamstress, an inventive knitter and a voracious reader"—and a chain-smoker, pun addict, and avid card player. Her dignified bearing and imposing stature (five-foot ten) matched her intense personality. A passage from William Wordsworth, the nineteenth-century English poet, sustained her through her trials:

> Knowing that Nature never did betray
> The heart that loved her.

It is ironic that just four years after Payne's initial foray into stellar compositions, Russell became the principal force in persuading astronomers of the overwhelming preponderance of hydrogen in the Sun and the stars. His own conversion followed more detailed observations of the Sun, and he rightly noted "a very gratifying agreement" between his findings about the Sun and Payne's earlier calculations for hotter stars. But he also left much unsaid. As Kidwell notes, "Russell . . . did not mention that Payne had dismissed her data on hydrogen as probably spurious, nor allude to his role in shaping this conclusion."

All doubts about the preponderance of hydrogen disappeared once the Sun's opacity was better understood and as others applied quantum mechanics to the problem. Today it is known that roughly 98 percent of the Sun's mass is made up of hydrogen and helium; all the heavier elements make up the remaining 2 percent. In number, hydrogen atoms dominate by far: for every thousand hydrogen atoms there are only sixty-three helium atoms. The next-most abundant elements, oxygen and carbon, contribute half an atom each for every thousand hydrogen atoms.

Russell's original suspicions in 1914 about cosmic abundances were, in the end, partly correct. Except for hydrogen and helium, the ratios among the heavy elements (sparse as they are) in the Sun do roughly match the ratios in the Earth. That is the signature of the common origin of the Sun and the planets out of a swirling cloud of matter some five billion years ago. Through Payne's pioneering efforts and the achievements of those who followed up on her suspicions a new understanding of the composition of the heavens arose: hydrogen became the dominant cosmic ingredient; earthly elements such as carbon, oxygen, nitrogen, and iron were just traces of "dirt" in the celestial mix. Nature, it seems, did not betray Payne after all.

CHAPTER FIFTEEN

Find a Way Around It

Stars, quasars, supernovae, galaxies—if it's
out of this world, she has seen it

O NE of the most important discoveries in modern astronomy was published in the *Reviews of Modern Physics* in 1957. Astronomers playfully refer to this paper as simply B^{2}FH (like some mathematical formula) from the initials of the authors' surnames. The first B refers to E. Margaret Burbidge, and the scientific journal article opens with a quotation from Shakespeare: "It is the stars, the stars above us, govern our conditions." The words, from *King Lear*, are certainly appropriate. With this paper, Burbidge and her colleagues—Geoffrey Burbidge (her husband), William Fowler, and Fred Hoyle—provided a map of the routes by which elements heavier than hydrogen and helium are forged within the fiery bellies of the stars. The calcium in our bones, the iron in our blood, and the oxygen we breathe all came from the ashes of ancient stars, which had either exploded as supernovae or died slowly, releasing their matter into space.

Margaret Burbidge's colleagues worked on theory and laboratory experiments; she employed the telescope, seeking

(Left to right) E. Margaret Burbidge, Geoffrey Burbidge,
William Fowler, and Fred Hoyle—the B²FH team in 1971.
(AIP Emilio Segrè Visual Archives, Clayton Collection)

direct evidence from the heavens. Others had speculated about the origins of elements, but B²FH delivered the proof. Stellar nucleosynthesis became a game of billiards around the periodic table. The balls are neutrons and protons, and the resulting elements serve as markers for the steplike evolution of a star.

When the bulk of hydrogen in a star's core is converted to helium, its central furnace will flame out, and nuclear burning will then take place in a shell of hydrogen surrounding the inert helium core. But eventually, as more and more helium is dumped upon the core, the helium at last ignites and fuses into carbon and oxygen. If a star is more massive than our Sun, the fusion process continues. The carbon core gets surrounded by a helium-burning shell, with a hydrogen-burning shell farther out. The center of the massive star starts developing a series of layers, akin to the structure of an onion. The carbon and oxygen atoms, heated to a billion degrees, go on to fuse into neon and magnesium. These, in turn, can serve as the raw materials in the construction of even heavier elements, such as silicon, sulfur, argon, and calcium, each chemical group burning (that is, fusing) in successive concentric shells.

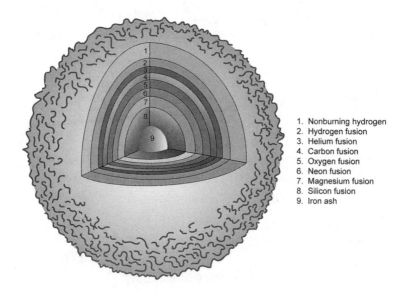

1. Nonburning hydrogen
2. Hydrogen fusion
3. Helium fusion
4. Carbon fusion
5. Oxygen fusion
6. Neon fusion
7. Magnesium fusion
8. Silicon fusion
9. Iron ash

Cutaway of a massive star, building up heavier elements in its interior as it ages.
(Illustration courtesy of Barbara Schoeberl, Animated Earth, LLC)

If a star is massive enough, the equivalent of eight or more Suns, its core will continue fusing elements until iron is formed, which is the end of the line. The fusion of elements heavier than iron requires more energy than it releases. That's when either a neutron star or black hole is born (as discussed in Chapter 7).

Burbidge first became aware of the stars in 1923 at the age of four. The young Eleanor Margaret Peachey beheld them with vivid clarity during a nighttime crossing of the English Channel. Upon starting her studies at University College, London, in 1936, she discovered to her delight that astronomy could be a career as well as a hobby. Wartime blackouts, which kept the urban sky dark, allowed her to carry out research on the outskirts of London, using a telescope so antiquated that it was moved by a hanging weight.

After the war, Peachey got married. She was eager to seek out the best telescopes and the best skies for observing, so she and her physics-trained husband, her closest collaborator over the years, moved to the United States. They eventually settled in California, where she began her lifelong commitment to opening up opportunities for women in science. At a time when women weren't allowed to use the telescopes (or even stay overnight in the observatory's dormitory), she fought for—and won—access to those atop Mount Wilson near Pasadena. "Thanks to her influence," Caltech astronomer Anneila Sargent has said, "women can observe at any American observatory." Burbidge puts it simply: "If you meet with a blockage, find a way around it."

Verifying that we are composed of stardust was the first in a long list of achievements. Burbidge made pioneering measurements of the masses of galaxies and specialized in quasars. These celestial entities are believed to be luminous objects at the centers of galaxies, where a spinning, supermassive black hole generates tremendous radiation. For many years she held the record for finding the most distant quasar, a feat listed in the *Guinness Book of Records*.

Burbidge returned briefly to Great Britain in 1972 to serve as the first female director of the famed Royal Greenwich Observatory. But, happier at a telescope than a desk, she soon returned to the University of California at San Diego, where she studied quasars that emit large amounts of X-rays.

She has not always embraced the majority opinion on celestial matters. Both Burbidges suspect that the true nature of quasars has not been fully revealed and that quasars are closer to us than most astronomers assume. In a 1994 memoir, she says that she is "continually surprised by the almost religious fervor with which most astronomers demand a single 'Big Bang' act of creation for the Universe." Burbidge is more attracted to the notion, introduced by Hoyle, that matter was created in successive epochs, not just by a single event. Her unconventional views have often spurred the astronomical community to new lines of research.

Burbidge has devoted more than eighty years to keeping watch on the universe. Unlike today's astronomers, most of whom sit in control rooms watching data displayed on monitors, she has had the pleasure of sitting directly at a telescope. "To ride with the telescope," she once recalled, "was an experience I wish I could share with today's generation. . . . One could look out at the spectacular vision of the heavens." She has held a front-row seat on a golden age of astronomy.

CHAPTER SIXTEEN

Dark Matters

Searching for the universe's main ingredient

EARLY half a mile beneath the surface of the Earth, within a cavern of an old iron-ore mine in northeastern Minnesota, special detectors cooled almost to absolute zero (−459.67 degrees Fahrenheit) are on the lookout. They serve the Cryogenic Dark Matter Search (CDMS), one of several projects around the world attempting to find a novel type of matter that has been long hypothesized but never seen. New particle physics theories, beyond the so-called standard model, suggest that all around us could be ghostly particles that blithely whiz through us with nary a nudge. The hope is that deep underground, far from disruptive cosmic rays, one of these exotic particles will occasionally bump into a detector and release an indisputable signal.

If and when that happens, astronomers will be jumping for joy. Along with opening up new physics, the discovery of such weakly interacting massive particles (or WIMPs) might solve a cosmic mystery that has endured for more than eighty years.

Those particles—distinct from those in the standard model, including the recently headlined Higgs boson—could be the long-sought "dark matter" thought to permeate the universe.

The first person to wonder about this unseen cosmic ingredient was an irascible physicist named Fritz Zwicky. A Bulgarian-born Swiss national, Zwicky arrived at Caltech in 1925 to study the properties of liquids and crystals. But that was just for starters. An aggressive and stubbornly opinionated man, he regularly annoyed his physics and astronomy colleagues by studying anything he pleased. Along the way he championed some pretty wild ideas, some of which proved their worth decades later. In 1933, as noted in an earlier chapter, he was the first to propose that a supernova—the total destruction of a star—left behind an extremely small and dense object that he called a "neutron star." The first such object wasn't detected until 1967.

Given his eclectic scientific style, it's not surprising that Zwicky also spied one of the first signs that the universe's ledger books were not quite balancing. He had decided to examine all the velocity information then available in the literature on the galaxies situated within the famous Coma cluster, a rich group of hundreds of galaxies some 330 million light-years distant. His statistical analysis revealed that the galaxies were moving around in the cluster at a fairly rapid clip. But adding up all the visible light being emitted by these galaxies, he realized that there was not enough luminous matter to bind the speeding objects to one another through the force of gravitation.

"It is difficult to understand why under these circumstances there are any great clusters of nebulae remaining in

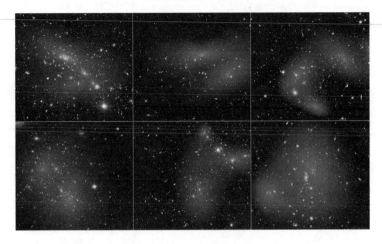

Dark matter was traced in these six different galaxy clusters and
depicted as mistlike clouds that surround each cluster.
(NASA, ESA, D. Harvey [École Polytechnique Fédérale de Lausanne,
Switzerland], R. Massey [Durham University, United Kingdom], the Hubble
SM4 ERO Team, ST-ECF, ESO, D. Coe [STScI], J. Merten [Heidelberg/
Bologna], HST Frontier Fields, Harald Ebeling [University of Hawaii at
Manoa], Jean-Paul Kneib [LAM], and Johan Richard [Caltech])

existence at all," he eventually concluded. The situation
seemed paradoxical. With the Coma galaxies buzzing around
so nimbly, the cluster should have broken apart long ago, but
it was still very much intact. Zwicky reasoned that some
kind of unseen matter must pervade the Coma cluster to provide additional gravitational glue. In his report to the Swiss
journal *Helvetica Physica Acta* in 1933, Zwicky referred to this
invisible substance as *dunkle Materie*, or dark matter.

Zwicky's suggestion was largely ignored for several decades. Astronomers at the time figured the dilemma would
disappear once they could analyze the motions of galaxies in

more detail. They presumed that "weighing" a cluster of galaxies would prove more complicated than Zwicky had supposed.

The issue wasn't revived until the 1970s, largely owing to Vera Rubin, an astronomer with the Carnegie Institution of Washington. Early in her career she had dabbled in quasar research, the study of the universe's most energetic galaxies. Having been recently discovered, quasars were then the hottest topic in astronomy, but Rubin came to dislike the field's cutthroat pace and so decided to seek a less stressful topic. She eventually turned her attention to a problem far less controversial, even boring: the rotation of spiral galaxies. In doing this, she teamed up with W. Kent Ford, who had recently perfected a new electronic instrument that made it easier to record the spectrum of a galaxy, the data needed to measure the rotation.

At this point, astronomers just assumed that a galaxy rotated much like our solar system, following the laws of gravity set down by Isaac Newton. The stars closest to the galaxy's massive center would travel faster than those farther out in the disk, where the gravitational influence is diminished—just as the inner planets in our solar system practically race around the Sun, while the outer planets move at a far slower pace. But in spiral galaxy after spiral galaxy, Rubin, Ford, and a team of Carnegie postdocs found a far different pattern. To their surprise, they revealed that the stars and gas at a disk's edge traveled just as fast as matter closer to the galaxy's center.

If the planets in our solar system acted like this, Jupiter, Uranus, and Neptune would have careered off into interstellar space long ago. Rubin recognized that a huge reservoir of

Fritz Zwicky coined the term
dunkle Materie (dark matter).
(Photograph by Fred Stein, courtesy of
the American Institute of Physics Emilio
Segrè Visual Archives.)

extra matter, imperceptible to her instruments, had to be tucked away somewhere to keep the stars from flying out of the galaxy. It was the Coma cluster problem all over again, but this time within an individual galaxy. Modeling this effect, theorists figured that each spiraling disk must be embedded in a large sphere of invisible matter to keep the luminous galaxy intact. They also knew that it couldn't be just ordinary matter that wasn't glowing, as the Big Bang didn't make enough regular particles to account for the dark matter required.

Some radio astronomers had measured a few of these fast galactic spins earlier, but by 1978 Rubin and her team had measured more than two hundred. This arsenal of data at last took the dark-matter problem off the back burner and turned it into one of the most active concerns in astronomy—an effort that continues to this day. While some astronomers initially questioned Rubin's findings, recent and more varied measurements have removed nearly all doubt.

Some of the best evidence to date is based on an effect known as "gravitational lensing." Astronomers, for example,

Vera Rubin measuring her spectra in the 1970s.
(AIP Emilio Segrè Visual Archives, Rubin Collection)

have aimed the Hubble Space Telescope at massive galaxy clusters to map their dark matter. While astronomers can't directly see the dark matter, they can view its gravitational effects, especially in the way it bends light arriving from the distant galaxies behind it, much like a lens. What results is an arty view of the cluster, filled with myriad arcs, bands, and rings of light (see chapter 17). The amount of light bending, using Einstein's rules of general relativity, provides the means to weigh the dark matter in the cluster and map its distribution.

On top of that, the exquisite measurements now made of the cosmic microwave background, the remnant radiation left over from the Big Bang, tell us that there is five times as much dark matter in the universe as there is of the ordinary elements that make up the stars, nebulae, and us. We're merely the icing on the cosmic cake. What this invisible stuff is remains one of astronomy's greatest mysteries, and yet the answer to dark matter's composition may not come from *out there*—the farthest recesses of space-time—but possibly from instruments that stand watch deep down in the Earth.

CHAPTER SEVENTEEN

Cosmic Funhouse

An amusing relativistic effect turns into
an important astronomical tool

T HERE'S always something delightful that catch-
es my eye after the Sun has set: on one evening
the artistic swoosh of a crescent moon, on another the striking pattern of stars that forms the
Orion constellation, whose appearance in the Northern
Hemisphere heralds the coming of winter. So, when looking
up at the nighttime sky I often smile.

And the cosmos, I have learned, is smiling back . . .
literally.

While searching through images collected by the Sloan
Digital Sky Survey, astronomers from Great Britain, Russia,
and Spain announced in 2009 that they had come across a
familiar face in the direction of the constellation Ursa Major—
that of the disappearing Cheshire Cat in Lewis Carroll's *Alice's
Adventures in Wonderland.* The two "eyes" of the cat are giant
elliptical galaxies, each the brightest member of a small group
of galaxies. Both groups are situated some 4.6 billion light

years away. More recently, NASA's Chandra X-ray Observatory discovered that these two sparse clusters are, in fact, racing toward one another at around 300,000 miles (480,000 kilometers) per hour and will eventually merge about one billion years from now.

But what's most captivating about this celestial formation is its "grin," a lustrous smirk generated by the two elliptical galaxies and their surrounding matter. As the light waves from

This group of galaxies has been nicknamed the "Cheshire Cat" because of its resemblance to the smiling feline in *Alice's Adventures in Wonderland.* The two "eyes" are elliptical galaxies, while the "grin" and "face" are formed by galaxies farther out, whose images are stretched out by gravitational lensing. This picture is a composite, blending an optical image with an X-ray image.
(X-ray: NASA/CXC/UA/J. Irwin et al.; Optical: NASA/STScI)

background galaxies farther out come upon the gravitational influence of all this matter in their journey through space, the distant light gets bent and stretched into long arcs. With a powerful enough telescope, you can see such smiles all over the celestial sky. The Cheshire Cat is only one of many examples of this funhouse effect that is fully explained by Einstein's general theory of relativity.

With his new gravitational theory, introduced in 1915, Einstein posited that space and time join up to form a palpable object, a sort of boundless rubber sheet (although in four dimensions). Masses, such as a star or planet, indent this flexible mat, curving space-time. With that image in mind, he predicted that a beam of starlight would noticeably shift as it passed by a massive celestial body, following the curved pathway. It was a prediction that thrust Einstein into the public eye: when astronomers, who were monitoring a 1919 solar eclipse, saw starlight graze the darkened Sun and get deflected by exactly the calculated amount, Einstein became world-famous overnight. "Lights All Askew in the Heavens," blared the headline in the *New York Times*, "but Nobody Need Worry."

In this situation, the Sun had become the gravitational equivalent of an optical lens. Instead of glass deflecting the light rays, gravity was doing the job. It wasn't long before others wondered whether such "gravitational lensing" might be sighted farther out. In 1920 the British astronomer Arthur Eddington considered the possibility of seeing multiple images of a star, if that star were properly situated behind another stellar body. Although the physical principle is not the same, you might think of the starlight as a stream of water that comes upon a rock and gets diverted into several streams on either side of the stone. Thus our eyes detect multiple images of the star, rather than just one. But in the end, Eddington figured that the effect would be so weak "as to make it impossible to detect it."

Four years later, the Russian physicist Orest Chwolson noted that if the distant star were aligned just right—precisely behind a star that acts as a gravitational lens—its light would spread out to form a ring that completely surrounds the lens. Einstein was already aware of these possibilities. As early as the spring of 1912, three years before he published his general theory of relativity, he carried out some calculations of gravitational lensing in his notebook and jotted down the possibility that a lens might not only create a double image of a star, but might also magnify the intensity of the star's light. However, he then dropped the subject.

Einstein didn't return to the problem until 1936, and then only after he was prodded by a young Czech electrical engineer and amateur scientist, who asked him to once again consider cosmic lensing. "Some time ago, [Rudi] W. Mandl paid me a visit and asked me to publish the results of a little calculation, which I had made at his request," wrote Einstein in his paper for the journal *Science* titled "Lens-Like Action of a Star by the Deviation of Light in the Gravitational Field." He went on to say it was "a most curious effect" but also concluded (like Eddington) that there was "no hope of observing this phenomenon directly," since it defied "the resolving power of our instruments." Privately, Einstein wrote the editor of *Science* that his findings had "little value, but it makes the poor guy [Mandl] happy."

But at the California Institute of Technology, Fritz Zwicky, both physicist and astronomer, thought otherwise.

The following year in *Physical Review* he pointed out that "extragalactic nebulae [galaxies] offer a much better chance than stars for the observation of gravitational lens effects." Acting like a giant magnifying glass, the galactic lens would enable astronomers to "see [other] nebulae at distances greater than those ordinarily reached by even the greatest telescopes," wrote Zwicky. It was a prescient vision, but one that was not confirmed for another forty-two years.

In 1979, British astronomer Dennis Walsh was closely examining a photographic plate to locate the visible counterpart of a newly discovered radio source, 0957+561, when he noticed that the radio object's position coincided with *two* star-like bodies, not just one. Additional telescopic observations from the Kitt Peak National Observatory in Arizona confirmed that the cozy pair were quasars. The spectra of these quasars were nearly identical, which hinted that they were not simply the chance alignment of two separate objects (which often happens). A celestial object's spectrum is as distinctive and exclusive as a fingerprint or personal sample of DNA. The spectral matchup strongly suggested that Walsh was seeing the *same* quasar—the brilliant core of a young galaxy some nine billion light years distant—but in duplicate.

Walsh and his colleagues reported their suspicion that a gravitational lens was at work, and further observations by other astronomers at the Palomar Observatory in California confirmed that conjecture. The lens turned out to be a giant elliptical galaxy, a member of a rich cluster of galaxies located halfway between the quasar and Earth.

It wasn't long before astronomers uncovered many other cases of gravitational lensing throughout the celestial sky—

and not just multiple images of pointlike objects. When entire galaxies are lensed by intervening galaxies or even clusters of galaxies, their broader shapes are often smeared into long arcs and rings. That's how the Cheshire Cat got its grin.

Gravitational lensing, however, has turned out to be far more than an amusing or pretty optical effect. Today it is one of astronomy's most valuable tools. The amount a light beam is deflected depends on the total mass of the gravitational lens. So, by carefully measuring the deflections, astronomers can "weigh" entire clusters of galaxies. Their results have confirmed that around 90 percent of the mass in these clusters is indeed composed of an unknown dark matter. Moreover, both the position and intensity of the arcs formed around the cluster of galaxies allow astronomers to map how this matter is distributed through and around the cluster. Such information is offering clues as to the true nature of dark matter.

And just as Zwicky forecast eight decades ago, gravitational lenses are magnifying the images of galaxies residing in the most distant regions of the universe, galaxies that would have been too small or faint to be seen with a telescope alone. All of these applications are helping astronomers trace the growth of galaxies and clusters of galaxies through time, to examine how cosmic structures have evolved and changed over the eons. "The vistas we uncover with this new gravitational telescope," writes astronomer Evalyn Gates in her book *Einstein's Telescope*, "will take us further than ever . . ., providing answers that may unlock the door into a deeper understanding of the fundamental nature of space, time, matter, and energy."

With improved technology, astronomers have also come to see individual celestial objects act as gravitational lenses, the enterprise that both Eddington and Einstein had deemed hopeless. Background stars in our Milky Way and in the Magellanic Clouds are seen to briefly magnify—microlens—due to dark objects passing in front of them. This is one way that astronomers have revealed the presence of both brown dwarf stars and extrasolar planets, objects too dim to be seen directly. Such an amazing accomplishment brings a smile to astronomers' faces—or even a broad, mischievous grin.

CHAPTER EIGHTEEN

Rivers of Galaxies

Once thought to be illusory, superclusters of
galaxies are now being well mapped

THE Milky Way has a new address. For more than
six decades it's been known that our galactic
home is perched at the edge of a long and vast
collection of galaxies called the Virgo Supercluster. But an international team of astronomers announced in
recent years that we belong to an even larger assembly in
this sector of the universe. Led by R. Brent Tully of the University of Hawaii at Manoa, the team dubbed this gargantuan
structure "Laniakea," which means "immense heaven" in
Hawaiian.

This finding proves, once again, that galaxies are very sociable creatures. Even though space-time is continually
stretching, moving most galaxies away from one another as
the universe expands, gravity keeps adjacent neighbors together, even drawing them closer, forming arrangements
across a range of sizes.

The Milky Way, for example, is part of a small collection
right here in our galactic neighborhood. Edwin Hubble named

it (rather uninspiredly) the "Local Group." One end is anchored by our home galaxy, surrounded by a bevy of dwarf galaxies; the Andromeda and Triangulum galaxies dominate the other end, with their own small companions. But the Local Group pales in comparison to the richest clusters. The Coma cluster, located some 300 million light-years away in the direction of the Coma Berenices constellation, contains thousands of galaxies hovering together like a dense cosmic flash mob.

Even before astronomers knew that many of the nebulae they were observing all over the heavens were distant

The Hubble Space Telescope imaged a large portion (several million light-years across) of the Coma cluster. The spherical cluster is more than 20 million light-years in diameter and contains thousands of elliptical and disklike galaxies.
(NASA/ESA/Hubble Heritage Team [STSci/AURA])

galaxies, they noticed how some of these cosmic clouds crowded together. In the eighteenth century, astronomer William Herschel wrote about Coma's "remarkable collection." Having built the largest telescopes in his time, he was able to spot this prominent swarm more than two centuries ago.

But how far did this tendency go? Were there also, astronomers asked, clusters of galaxy clusters? That question took quite a while to answer. In the 1930s, both Harvard astronomer Harlow Shapley and the Swedish astronomer Erik Holmberg spoke of "metagalactic systems" or "metagalactic clouds," what we today call superclusters. To these observers' eyes, some of the clusters appeared to form even larger assemblies.

But, around the same time, Hubble photographed selected regions of the sky and concluded the opposite: that clusters were distributed fairly uniformly across the heavens. Hubble at the time was embracing the cosmological principle, the idea that on the very largest scales the universe must be "isotropic"—looking about the same no matter in which direction you looked. To him, galactic groupings stopped at clusters. This view was so strong that few dared to question it, and Hubble's opinion prevailed for many years . . . until a feisty French astronomer began to alter that widely held belief.

During World War II in France, astronomer Gérard de Vaucouleurs had been an expert observer of Mars, but by the early 1950s he had traveled to Australia to work at the Mount Stromlo Observatory. There he performed a tedious yet very important chore: a revision of one of astronomy's bibles, the Shapley-Ames catalog of bright galaxies. It changed his professional life. While updating the catalog's listings to include Southern Hemisphere galaxies, he couldn't help but notice (with the aid of his telescope) that the Milky Way, along with its Local Group neighbors, is caught on the outskirts of a much larger system of galaxies. Altogether this system is generally arranged as a flat disk, made up of multiple clusters of galaxies. On a celestial map, it appears as a long band that stretches across both the northern and southern skies. The Virgo cluster, a huge collection of hundreds of galaxies located some 65 million light-years away, serves as the disk's centerpiece.

De Vaucouleurs was seeing what Holmberg and Shapley had already noticed, but he was more tenacious. In a 1953 scientific paper, he gave this grouping a distinct name. He called it the "Local Supergalaxy," what later became known as either the Local or Virgo Supercluster. In the 1980s de Vaucouleurs recalled that his suggestion was largely received with resounding silence. "It was considered as sheer speculation, even nonsense," he told me. "Some prominent astronomers even told their students that it was an insane topic to work on. The concept that the universe was isotropic was too strong. It was dogma."

But a few listened and gradually examined the idea further. More and more evidence piled up as other astronomers began to carry out their own surveys of galaxies across the heavens, with new instrumentation that enabled them to find both nearby and distant clusters that were once too faint to be counted. "All of a sudden," wrote Italian astronomer Andrea Biviano in a review of this history, "researchers had a catalogue of clusters, and they could start to look at them as a population, rather than as individual objects."

By 1961 the Virgo Supercluster was not alone. That year UCLA astronomer George Abell, the most noted cluster hunter of his era, examined all the data gathered so far by both him and others and pointed out other potential superclusters, each "large cloud" stretching up to 160 million lightyears from end to end. Abell counted seventeen more nearby in our universe. As for the Virgo Supercluster, Abell declared that an independent survey found "striking confirmation of de Vaucouleurs' hypothesis."

But acceptance did not come readily. No one could yet explain how such large structures could remain stable over the eons. More than that, some astronomers wondered if they were being deceived. Our eyes are very sensitive to patterns, a trait that enabled our ancestors to spot a predator amid the jungle foliage. For a long time, many were wary that supercluster proponents were merely tracing out shapes in a random distribution of clusters, much the way early planetary astronomers found "canals" on Mars.

Starting in the 1980s, however, as astronomers were able to determine the distances to more and more galaxies and clusters, they produced three-dimensional maps of the heavens. They discovered they weren't being fooled at all. In fact, the distribution of galaxies was more astounding than they had ever imagined. Galaxies appear to congregate as if they are on the surfaces of huge, nested bubbles, with the bubble interiors nearly devoid of galaxies. Evidence suggests that this cosmic foam originated in the Big Bang, owing to perturbations surging through the primordial soup.

Filamentary superclusters stand out where the bubblelike surfaces intersect. At the time of de Vaucouleurs's death in 1995, many of these superclusters were well mapped, with astronomers naming them after the constellations in which

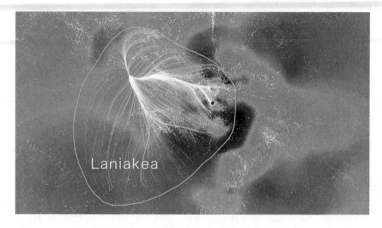

The enclosed line circles the Laniakea structure. The Milky Way galaxy (located at black dot on right) is on the edge of the Virgo Supercluster streaming inward.
(Copyright © 2014, Nature Publishing Group)

they can be found, such as Coma, Leo, Hercules, Perseus-Pisces, and Centaurus.

And these superclusters are not static. That's how Tully and his colleagues found Laniakea. They saw that the Virgo Supercluster is being gravitationally drawn, like a river flowing downhill into a larger sea, toward a dense collection of galaxies known as the "Great Attractor." By tracing the movement of galaxies directed toward the Great Attractor, they could define the borders of the new Laniakea Supercluster.

Home to some 100,000 galaxies, Laniakea stretches more than 500 million light-years across, nearly five times larger than our original Virgo abode, which is now a mere branch. Formerly caught in a supercluster suburb, the Milky Way finds itself in Laniakea's hinterlands.

CHAPTER NINETEEN

The Big Dipper Is Crying

The well-known constellation looks as if it is leaking cosmic rays

O NE year after it opened in 2007, the Northern Hemisphere's largest cosmic-ray detector, the Telescope Array situated in western Utah, began observing a relatively large number of ultra-high-energy cosmic rays emanating from just below the handle of the Big Dipper. What might be the exact cause of this "hotspot" of rays? Even after years of observation and study, no one knows for sure. "All we see is a blob in the sky," says University of Utah astrophysicist Gordon B. Thomson, "and inside this blob there is all sorts of stuff—various types of objects—that could be the source."

An international team of astronomers found the hotspot by tracking a cascade of secondary particles that showered down upon the Earth and were captured by the Telescope Array when particularly powerful cosmic rays—those above 57 billion billion electron volts (14 million times the energy of the particles accelerated recently in the Large Hadron Collider)—hit the atmosphere. The array, a high-tech wonder, consists of more than five hundred scintillation detectors, each about the size of a Ping-Pong table, spread out over three hundred square miles of desert like myriad chess pieces. These measure the secondary particles that rain down upon the surface when a cosmic-ray shower hits the Earth's atmosphere. Positioned around these detectors are three stations, each with a set of mirrors watching for blue flashes also created by the incoming cosmic rays. It's a modern-day method that's a far cry from the cruder instruments used by the discoverers of cosmic rays a century ago.

At the start of the twentieth century, researchers were only just discovering that charged ions reside in the air. Did

In this time-lapse photo, stars appear to rotate above a section of Utah's Telescope Array, which is aimed at detecting highly energetic cosmic rays from space.
(Ben Stokes, University of Utah)

this ionization originate from the Earth's crust, they wondered, or from radioactivity within the atmosphere itself? Or perhaps from even farther out, the atoms getting ionized by some type of radiation journeying from the Sun?

Fascinated by this mystery, Theodor Wulf, a German priest and physicist, built a sensitive electroscope (the era's standard charge detector consisting of wires or metal leaves suspended in a vessel) and, while on a trip to Paris in 1910, took his new instrument to the top of the Eiffel Tower, then the world's tallest structure. Figuring the radiation emanated

Victor Hess (center) in 1911 about to depart on
an air-balloon flight from Vienna.
(*New York Times*)

from the ground, he expected to measure a far weaker signal nearly one thousand feet (300 meters) above the cityscape. But instead, his signal was surprisingly strong. Perhaps, Wulf mused, it was coming from radioactive iron within the tower's ornate lattice.

Continuing this quest, a number of scientists started taking measurements aboard balloons, which could reach higher altitudes, but their results were contradictory. It was Austria's Victor Francis Hess who gathered the first convincing evidence that the radiation was arriving from outer space. Hess, an ardent amateur balloonist, was a physicist at the newly opened Institute for Radium Research in Vienna. His moment of discovery came on August 7, 1912, after he and two companions took off in a hydrogen-filled balloon from the Bohemian town of Aussig for the seventh in a series of flights he had been conducting that year. Using three electrometers of improved accuracy, he detected a noticeable increase in his ionization readings as his balloon rose to an altitude of 3.3 miles (5.3 kilometers). In fact, the ionization was three times higher than on the ground. Hess knew he was too far up for this radiation to be arriving from below. That meant it must be *Höhenstrahlung*, as he called it, "radiation coming from above."

This was not a eureka moment for the scientific community, however. Many were still skeptical, including the world-renowned Caltech physicist Robert A. Millikan, who in the 1920s used unmanned balloons to take his instruments to even greater heights, up to nine miles (fourteen-and-a-half kilometers). As late as 1924 he reported that "the whole of the penetrating radiation is of local origin." But after continuing his measurements atop mountains and aboard airplanes, he was at last convinced of their extraterrestrial nature. Millikan, a bit of a showboater, didn't mind that America's newspapers gave him all the credit for the find, with no mention of Hess.

Millikan has "found wild rays more powerful and penetrating than any that have been domesticated or terrestrialized . . . probably completing [an] alphabet for the language by which the stars communicate with man," reported the *New York Times* on November 12, 1925. "The mere discovery of these rays is a triumph of the human Mind that should be acclaimed among the capital events of these days."

Millikan, like many others at the time, believed the radiation was electromagnetic in nature. Because the radiation was so penetrating, he figured the wavelengths had to be shorter than gamma rays. At a meeting of the National Academy of Sciences, he called them "cosmic rays." With great imagination, he declared that the highly energetic photons were released when particles in interstellar space somehow condensed into higher elements. To Millikan, cosmic rays were the "signals broadcasted throughout the heavens of the births of the common elements . . . the birth-cries of the infant atoms."

This led to a raging battle between Millikan and University of Chicago physicist Arthur H. Compton, who was sure that the interstellar "rays" were actually particles. The debate between the two was so fierce that the national press regularly covered this scientific tussle. The particle model finally won in 1932, once Compton sent teams of researchers around the globe, from Alaska to New Zealand, and fully demonstrated that the rays varied in intensity with latitude. The cosmic rays increased in number as the researchers traveled from the equator to the poles. That meant they were particles getting deflected by the Earth's magnetic field: the field lines point toward the poles, and particles swoop more easily toward the polar regions than equatorial latitudes. (Photons are not diverted by magnetic fields.) With the controversy settled over the rays' true nature, full credit was also restored. It was Hess who was awarded the 1936 Nobel Prize for his original discovery more than two decades earlier. The fact that everyone continued to call the alien particles cosmic rays was Millikan's consolation prize.

As the use of Geiger counters and cloud chambers grew more sophisticated, physicists came to see that cosmic rays were mostly protons, but could also be atomic nuclei or electrons. They enter the Earth's atmosphere, in a range of energies, from all directions of the celestial sky. Some five quintillion (5×10^{18}) strike the Earth's atmosphere each second. Upon colliding with air molecules, the primary rays generate a cascade of secondary particles that plummet to the ground (and get detected by such instruments as Utah's Telescope Array).

Cosmic rays gave birth to the field of particle physics. By carefully studying cosmic-ray interactions, physicists came to discover new and bizarre elementary particles, beyond the plain-vanilla proton, electron, and neutron. In 1932 the positron (the electron's antimatter mate) was discovered in a cosmic-ray cloud chamber; by 1937 the track of a speeding muon (a heavy electron) was similarly spotted in a chamber photograph.

By the 1950s, with particle physicists constructing big accelerators to search for new particles, cosmic-ray physicists began to focus more on the origin of the "rays." How and where are cosmic rays being created in the vastness of the universe? they asked. Millikan was wrong on all counts. But other ideas were already in the wind. As early as 1934, Walter Baade of the Mount Wilson Observatory and Caltech physicist Fritz Zwicky suggested that the rays came from spectacular stellar blasts, explosions earlier dubbed "supernovae." It's an idea that holds up to this day. More recently, active galactic nuclei have also come to be suspected as rich sources of the rays: a spinning supermassive black hole at a galaxy's center spews out blazing jets of particles into space, acting like an electrical generator as it rotates.

All of the above may be contributing to the signal gleaned by the Utah array. The Milky Way lives in the outskirts of the Virgo Supercluster of galaxies, and the hotspot resides in the very direction of that vast supercluster, home to tens of thousands of galaxies. The cosmic rays arriving on Earth from that bearing could then be the collective shout from the myriad supernovae and active galaxies occupying the supercluster. Here we are, they are saying, here we are.

CHAPTER TWENTY

Einstein's Symphony

Finding Einstein's long-sought ripples in space-time

EADING in from the southern sky at the speed of light, a gravitational wave passed through the Earth on September 14, 2015, in less than a second. Such events have occurred ever since our solar system coalesced out of a nebulous cloud more than four billion years ago. But this time was different. This time researchers finally snared that faint swell in space-time, ushering in a new age of astronomy as game-changing as the telescopic era introduced by Galileo.

Einstein first mentioned the possibility of gravitational waves (or gravity waves, as they're more popularly known) more than one hundred years ago. He predicted that a pair of masses, such as two stars moving around each other, would undulate the very fabric of space-time. These waves then would move outward, much like the ripples generated when a stone is dropped into a pond, getting weaker and weaker as they spread. This pattern is far different from the way electromagnetic waves propagate. Light travels *through* space;

gravity waves, by contrast, are vibrations *in* the very framework of space-time—compressing and stretching space-time (and any object caught within it) as they pass by.

Ever since the 1960s, scientists had been attempting to capture a gravity wave. At the University of Maryland, physicist Joseph Weber constructed the first detectors, large cylinders of metal, called Weber bars, surrounded with sensors that he configured to "ring" like a bell whenever a gravity wave passed through them. He claimed to have observed such ringing a number of times, starting in 1969, but the detections were never confirmed. His effort, however, founded a new field of study, stimulating others to come up with new schemes.

In 1972, at MIT, physicist Rainer Weiss wrote a landmark report, the first complete examination of an approach known as "laser interferometry." He suggested arranging a set of mirrors in the form of an L—one in the corner, the others at each end; continually bouncing laser beams up and down each arm to keep an accurate tab on the distance between them; and then having the beams recombine (optically "interfere" with one another) to check if a gravity wave had wiggled the mirrors.

Weiss and others built small laboratory prototypes, but the MIT physicist knew that no cosmic waves would ever be found unless the mirrors were separated by miles. The longer the distance, the greater the sensitivity of the measurement. By the 1980s, tired of his slow progress, Weiss joined forces with Caltech theorist Kip Thorne, then the world's top expert on gravity waves, and experimentalist Ronald Drever, also at Caltech, to take a giant leap and seek National Science

Foundation (NSF) funds to construct a pair of large detectors with arms two-and-a-half miles (four kilometers) long, set geographically apart to rule out local noise.

Upon hearing of this proposal, the physics community quickly protested; it was aghast at the idea that the NSF might spend money on such a gamble when so much of the technology still needed to be invented. It was only after a decade of campaigning and politicking that the funds were finally approved and ground was broken for the Laser Interferometer Gravitational-Wave Observatory (LIGO) in 1994. One detector resides in Livingston, Louisiana, the other 1,900 miles (around 3,000 kilometers) northwest in Hanford, Washington. Turned on in 2001, and advanced and improved over the

This illustration shows the generation of gravitational waves, ripples in space-time, as two black holes spiral into one another, heading toward eventual collision.
(LIGO / T. Pyle)

years, both instruments at last found their quarry that fateful September night.

The wave first arrived at Livingston two hours before dawn, at exactly 4:50:45 a.m. Central Daylight Time. Seven-thousandths of a second later, Hanford also sensed the wave. But the operators in the main control room at each site didn't notice. LIGO was then conducting an engineering run, a check on some newly installed equipment. Data was being collected, but the sound alert, which goes off whenever a candidate signal passes a certain threshold, was not on. That awaited the official scientific run of the new, "advanced" detectors, which was set to occur a few days later.

Instead, the data silently streamed into the automatic analysis pipeline, where, within a few minutes, the waveform popped up on the computer monitor of LIGO collaborator Marco Drago at the Albert Einstein Institute in Hannover, Germany. A member of LIGO's coherent wave burst group, the young postdoc was among the first to see the signal. It was beautiful, clear, and strong. In fact, it was so picture-perfect that Drago and his colleagues, who soon gathered together, just assumed it was a "blind injection," someone from LIGO secretly sending out a fake signal to test the system. But they soon learned that wasn't the case. Could it have been a hacker? That, too, was a concern and, therefore, was thoroughly checked out. In the end, LIGO scientists finally realized they had their Cinderella scenario or "golden event," as Drago put it—a gravity wave always hoped for but never expected as a first detection. It stood high above the noise.

As LIGO continued to gather data, teams of theorists deciphered the inaugural wave's message according to Einstein's general theory of relativity. In less than a second, the signal had swept upward in frequency from about 30 hertz (cycles per second) to around nearly 300 hertz. Because that's the same frequency range as sound, it can be heard as a musical glissando that starts as a deep bass and swiftly ends near middle C. Gravity-wave astronomy is adding sound to our cosmic senses. This "chirp" was just the type of signal that would be expected to occur when two black holes, long orbiting one another, swirled together ever faster until they merged to form a single black hole. Such a collision had never before been demonstrated; the LIGO observations not only confirmed that it had occurred, but also indicated the sizes of the black holes. One of the holes weighed thirty-six solar masses, the other twenty-nine solar masses. The resulting combined black hole, at sixty-two solar masses, was less massive than the sum of the two because some of the mass was instantly converted into pure gravitational-wave energy—fifty times more energy than all the stars in the universe were radiating at that moment. At the collision site, such a spacequake would be deadly, but by the time the waves reached Earth some 1.3 billion years later, they moved the LIGO mirrors a mere fraction of the width of a proton. That's why only gravity waves from the universe's most violent events are currently measurable.

And this detection was just the start. Other signals were soon spotted in the ensuing months and years. The LIGO instrumentation is continually being improved, so that it will eventually be able to register waves arriving from even farther regions of the universe. A similar detector called Virgo is now operating in Italy in coordination with LIGO. Gravity-wave astronomers expect someday to see events weekly, possibly

The Laser Interferometer Gravitational-Wave Observatory in
Livingston, Louisiana.
(Caltech/MIT/LIGO Lab)

even daily. Black-hole collisions are their big game, but other
types of events are also expected to turn up. Kip Thorne describes them as "the warped side of the universe."

It was almost guaranteed that researchers would hear the
resounding crash of two city-sized neutron stars (paired together in a binary system) spiraling into each other as their
orbital dance decays. And in due course, they did. Both LIGO
and Virgo detected their first neutron-star collision on August 17, 2017. Such events may turn out to be the bread-and-
butter of these detectors' trade—and the most entertaining.
Less dense than black holes, a pair of neutron stars takes longer to merge, so the final recordable signal can last a minute
or more instead of fractions of a second. The gravity-wave

"telescopes" register a sinusoidal tune that sweeps to higher and higher frequencies as the two balls of pure neutrons spiral into one another. As soon as they touch, the two stars are shredded to pieces, releasing a burst of electromagnetic radiation across the spectrum, from radio waves to gamma rays. What happens after the collision depends on the situation: The remnants might coalesce into a new, more massive neutron star, if it's rotating particularly fast. Or if heavy enough, they might condense to utter invisibility, forging a black hole.

There will be another type of signal in the gravity-wave sky, although far less frequent. A solitary tsunami of a wave may hit our shores whenever a star within our local galactic neighborhood explodes as a brilliant supernova. This happens when the star's nuclear core runs out of fuel, collapses, and sends out a shock wave and a flood of neutrinos that blows the rest of the star apart. Examining the gravitational waveforms from such a spectacular event will allow astronomers to see, for the first time, the birth of a neutron star or black hole at the end of a star's life.

All the while, playing in the background amid these chirps and pops, could be ongoing rhythms—a steady hum. When a neutron star forms, for instance, it might briefly vibrate and develop a bump on its surface, an inch-high "mountain" that freezes into place for a while. This deformation, jutting out like a finger, would send out a continual set of gravity waves as it continually "scrapes" the space around it.

And beneath all those varied gravity-wave songs, astronomers expect an underlying murmur—constant, unvarying, and as delicate as a whisper. This buzz would be the faint reverberation of our universe's creation, its remnant thunder echoing down the passages of time. "That is the prize," says MIT physicist Nergis Mavalvala. That's because these primordial waves would bring us the closest ever to our origins, perhaps verifying that the universe emerged as a sort of quantum fluctuation out of nothingness. Future laser interferometers in space may be the first to see this gravitational-wave background.

Finally, there is the tantalizing prospect of encountering the unanticipated. Some theorists already wonder whether there might be relics from the early universe, highly energetic "defects" that were generated as the cosmos cooled down over its first second of existence. These include one-dimensional cosmic strings, extremely thin tubes of space-time in which the energetic conditions of the primeval fireball still prevail. Wiggling around like rubber bands, they would produce plenty of gravity waves. Not until astronomers scanned the heavens with radio telescopes did they discover pulsars and quasars. What else might be skulking about in the darkness of space, as yet unseen?

CHAPTER TWENTY-ONE

Underground Astronomy

Learning about the cosmos with detectors
buried in the Earth

DEEP beneath the South Pole, thousands of
detectors, set within a cubic kilometer of ice, lie
in wait. While looking up toward the surface,
they also peer downward, hoping to catch certain elementary particles from the northern sky that travel
through the Earth daily. Nearly all of these elusive particles—
called neutrinos—blithely pass through our dense planet like
ghosts on the run. Most of the time no signal is registered
by the instruments. But on rare occasions a neutrino and a
detector collide.

Between 2010 and 2013, this frigid array of detectors,
known as the IceCube Neutrino Observatory, recorded some
35,000 neutrinos journeying through our entire planet to the
Antarctic ice—a minuscule number compared with the trillions that traversed the Earth over that time. Most of the recorded neutrinos were generated locally, when cosmic rays
impacted the northern atmosphere. But a tiny fraction of them
appeared to have arrived from events far outside the Milky

An image of one of the highest-energy neutrino events registered by the IceCube Neutrino Observatory, shown at the bottom, superimposed on a view of the laboratory at the South Pole. When the neutrinos cross the underground detectors, they leave these tracks of light.
(IceCube Collaboration)

Way—either from massive stars exploding in distant reaches of the universe, or from the active cores of blazing galaxies. The ultrahigh energy of this special set of particles, far beyond the levels of the other neutrinos, revealed them for what they were.

With this success at identification, the IceCube detectors offer an entirely new way to survey the cosmos, an endeavor that couldn't have been imagined less than a century ago. Indeed, the very idea of the neutrino was first thought too crazy to be true, the physics equivalent of unicorns or elves. Even more peculiar was where the neutrino's story began: in a German prisoner-of-war camp during World War I.

The British physicist James Chadwick had been studying the phenomenon of radioactivity in Berlin under Hans Geiger (of Geiger counter fame) when the war broke out. Chadwick was soon sent to an internment camp set up at a racecourse just outside the city. To while away the hours of confinement, he began teaching physics to his prison-mate Charles Ellis, a young and sociable cadet from Great Britain's Royal Military Academy who had arrived in Germany on holiday just before the war's unexpected eruption. Together, the two compatriots organized a small research lab in one of the horse stables, an endeavor that was surprisingly tolerated by the camp's senior officials and generously supported by Chadwick's former German scientific colleagues.

The experience hooked Ellis. After the war, he committed to a career in physics instead of the army and ended up conducting experiments at the famous Cavendish Laboratory in Great Britain, where he studied a troubling anomaly. Whenever a radioactive nucleus decayed by ejecting an electron, something went awry. Ellis and a colleague noticed that the energy of the nucleus before it radioactively decayed was more than the total energy of the system afterward (that is, the combined energy of the depleted nucleus and the fleeing electron). It looked as if energy were disappearing, which violated one of the most sacred rules of physics—conservation of energy. Energy can be neither created nor destroyed.

But Wolfgang Pauli, a Viennese physicist, had an abiding faith that atoms were obeying the physical laws of the land, which led him to a radical proposition. In 1930, he suggested that an entirely new particle, invisible to ordinary instruments, could explain the energy discrepancy. Every time a nucleus spewed out an electron, it also emitted a neutral, phantom-like particle that seemed to vanish, carrying away that extra bit of energy and balancing the books.

Usually undaunted by new concepts, Pauli this time was intimidated by the outrageousness of his idea. "Dear radioactive ladies and gentlemen," he teasingly wrote his friends, then attending a physics conference in Germany. "For the

Wolfgang Pauli in 1930, around the time
he first hypothesized a new neutral particle,
later dubbed the neutrino.
(Photograph by Francis Simon, courtesy AIP Emilio
Segrè Visual Archives, Francis Simon Collection)

time being, I dare not publish anything about this idea and address myself confidentially first to you, dear radioactive ones, with the question of how it would be with the experimental proof of such a [particle]." He thought of his remedy as "desperate." It wasn't traditionally acceptable for theorists to conjure up particles out of whole cloth, especially particles that seemed impossible to catch.

That all changed in 1932. That year, Chadwick discovered the first known electrically chargeless particle—the neutron—which at last gave Pauli the courage to officially publish his idea that another neutral particle might exist. Soon after, physicist Enrico Fermi dubbed Pauli's hypothetical particle the neutrino, Italian for "little neutral one." The name was apt, for at the time the neutrino was thought to have no mass. According to Pauli's theory, it was nothing more than a spot of energy that flew off at the speed of light.

Despite Chadwick's discovery of the neutron, it took years to prove that neutrinos were more than figments of Pauli's imagination—so long, in fact, that some physicists began to call his particle "the little one who was not there." Pauli had reason to be apprehensive. The neutrino is so oblivious to ordinary matter that it would take a stack of lead, thousands of light-years in length, to stop one in its tracks. Neutrinos bolt through the Earth as if it's no more substantial than a cloudy mist.

But the odds of catching one are considerably increased if there is a flood of such particles coming at you. Indeed, that's how they were finally cornered. In the mid-1950s, physicists Clyde Cowan and Frederick Reines set up a detector outside a South Carolina nuclear power plant and each hour caught a few neutrinos out of the trillions generated by the reactor's core. Receiving news of the verification while attending a conference in Zurich, Pauli celebrated with colleagues by climbing the town's local mountain and enjoying several wine toasts at the top. With a friend on each arm helping him on the way down, Pauli turned to one and remarked, "All good things come to the man who is patient."

About a decade later, physicist Raymond Davis set up the first neutrino observatory in a gold mine, nearly a mile beneath the Black Hills of South Dakota. An underground location assured the measurements would be free from disruptive cosmic rays. In continuous operation for a few decades, Davis's detector kept watch on the torrent of neutrinos flung into the solar system as the Sun burned its nuclear fuel. It provided the first hint that the neutrino had a smidgen of mass after all. More advanced underground observatories constructed in the 1990s provided the ultimate proof, a confirmation that won the lead researchers for the experiments—Takaaki Kajita at the Super-Kamiokande detector in Japan and Arthur McDonald at the Sudbury Neutrino Observatory in Canada—the Nobel Prize in Physics in 2015.

Neutrino detectors and observatories can now be found or are under construction around the globe: not only in Antarctica, Japan, and Canada, but also in France, Russia, Italy, and India. And they are beginning to extend their searches beyond the neutrinos emanating from the Sun to the more powerful particles trekking through the cosmos. While weighing less than a billionth of the mass of a proton, each captured neutrino will help scientists understand the universe's history, structure, and future fate.

CHAPTER TWENTY-TWO

Eavesdropping on the Universe
The Galileo of radio astronomy

EVEN at fifty-five miles per hour, the desolate terrain seems to pass by in slow motion. Only an occasional stand of piñon pines on the side of a hill or, farther off, the stark profile of an erosion-sculpted mountain breaks the monotony.

But suddenly, after driving over a rise on Route 60, a few dozen miles west of Socorro, New Mexico, the weary traveler comes upon a spectacular sight: in the distance are twenty-seven dishlike antennas, lined up for miles over the flat, desert Plains of San Agustin. Airline pilots who fly over the ancient, mile-high lakebed have long called this gigantic Y-shaped installation "the mushroom patch." But ever since this facility was first dedicated in 1980, astronomers have simply referred to it as the VLA, for Very Large Array, one of radio astronomy's premier eyes on the universe.

Its majestic white dishes move in unison, like a mechanical version of the Rockettes, New York City's legendary dance company, to gather the radio waves sent out by myriad celestial objects. On one day, the antennas might trace the wispy outlines of a gaseous nebula to see how its molecules tumble and collide, leading astronomers to the birthplace of new stars. The next day, the dishes could point toward a supernova and snap a "radio picture" of the debris racing away from the monstrous explosion.

The array's particular strength is acting like a giant zoom lens. For a few months at a time, the antennas are crowded close, each arm of the Y no more than half a mile long. This setup provides a sort of wide-angle view of the heavens, perhaps to trace the gas clouds in a nearby galaxy. But to get a closer look, the antennas are periodically transported along railroad tracks out to greater distances, up to thirteen miles (twenty-one kilometers) along each arm. In the most extended

The Karl G. Jansky Very Large Array.
(Courtesy of NRAO/AUI)

arrangement, as the Earth sweeps the antennas around, the individual dishes collectively simulate the capability of a single antenna spanning some twenty-two miles (thirty-five kilometers), roughly the size of Dallas, Texas.

Over the array's nearly four decades of service, thousands of scientists from around the world have used the VLA to study the cosmos. Sometimes their focus is near—within our own solar neighborhood—and at other times out to the farthest reaches of space-time. On one occasion, visitors from Hollywood even took their turn: in the 1997 movie *Contact*, a fictional astronomer played by actress Jodie Foster used the iconic scopes to find radio proof for the existence of intelligent extraterrestrials.

But by the 1990s the National Radio Astronomy Observatory (NRAO), which operates the New Mexico array, recognized that the facility was getting long in the tooth, hindered by its 1970s-vintage electronics. So, in partnership with Canada and Mexico, the NRAO spent a decade upgrading the array's technology—from installing state-of-the-art receivers and fiber-optic transmission lines, to obtaining an innovative supercomputer to swiftly correlate its data. Ever since its completion in 2012, the new array has been detecting signals more than ten times fainter than the original system and is covering a radio-frequency range three times as wide, making it "by far the most sensitive such radio telescope in the world," says former NRAO director Fred K. Y. Lo. Need to take a cellphone call from Jupiter, some half a billion miles away? The new array can do it.

Given this transforming reincarnation, the NRAO decided it was also time to update the VLA's humdrum name, and so solicited suggestions via the internet from both the public and the scientific community. Candidate names flooded in from 17,023 people in more than sixty-five countries. Sifting through some 16,000 unique names, NRAO officials at last chose a new moniker that was eminently suitable. At a rededication ceremony that took place on March 31, 2012, the New Mexico facility was formally renamed the Karl G. Jansky Very Large Array.

Although hardly a household name, Karl Jansky is a pioneering giant to radio astronomers. He's the Galileo of radio astronomy. In the 1930s, Jansky set up a unique radio receiver amid central New Jersey's potato fields, and with it became the first to wrench astronomy away from its dependence on the optical spectrum, beyond the narrow hand of electromagnetic radiation visible to the human eye. His first, provisional step ultimately led to a new and golden age of astronomy that thrives to this day. But, as is often the case in astronomical history, Jansky began his investigations for a totally different reason.

In 1928, fresh out of college with a degree in physics and newly hired by Bell Telephone Laboratories, the twenty-two-year-old was assigned to investigate long-radio-wave static that was disrupting transatlantic radio-telephone communications. To track down the sources, he eventually built a steerable antenna—a spindly network of brass pipes hung over a wooden frame that rolled around on Model-T Ford wheels. It was known around the lab as "Jansky's merry-go-round."

Setting up his antenna near Bell's Holmdel station, Jansky soon learned that thunderstorms were a major cause of the disruptive clicks and pops during a radio phone call. But there

Karl Jansky with his "merry-go-round," the historic radio antenna
that initiated the field of radio astronomy.
(Reused with permission of Nokia Corporation)

was a steady yet weaker hiss that he also kept receiving. After
a year of detective work, Jansky at last established in 1932 that
the disruptive 20-megahertz static (a frequency between the
United States AM and FM bands) didn't originate in the
Earth's atmosphere, or on the Sun, or from anywhere within
our solar system. To his surprise, he saw that it was coming
from the direction of the Sagittarius constellation, where the
center of our home galaxy, the Milky Way, is located. Jansky
affectionately dubbed the signal his "star noise." For Jansky it
hinted at processes going on in the galactic core, some 26,000
light-years distant, that were not revealed by visible light rays

emanating from that region. For unlike visible light, radio waves are able to cut through the intervening celestial gas and dust, in the manner of a radar signal passing through a fog.

Jansky's unexpected discovery made front-page headlines in the *New York Times* on May 5, 1933, with readers being reassured that the galactic radio waves were not the "result of some form of intelligence striving for intra-galactic communication." Ten days later NBC's public affairs–oriented Blue Network broadcast the signal across the United States for the radio audience to hear. One reporter said it "sounded like steam escaping from a radiator."

By 1935, Jansky speculated that the cosmic static was coming either from the huge number of stars in that region or from "some sort of thermal agitation of charged particles," which was closer to the truth. Years later, astronomers confirmed that the noise was being emitted by violent streams of electrons spiraling about in the magnetic fields of our galaxy. Just as an electric current, oscillating back and forth within an earthbound broadcast antenna, releases waves of radio energy into the air, these energetic particles broadcast radio waves out into the cosmos. And Jansky was the first to detect them. He was Earth's first eavesdropper on the universe.

Despite the worldwide publicity, however, few astronomers then appreciated Jansky's new "ear" on the universe. Most were more comfortable with lenses and mirrors than with radio receivers. It was not until after World War II, spurred by the military development of radar technology, that the infant field at last took off. During the subsequent decades, radio telescopes were mapping the locations of colossal clouds of gas over the breadth of the Milky Way, discovering the existence of neutron stars from their metronomic radio "beeping," and helping astronomers unmask quasars as the violent cores of newborn galaxies in the distant cosmos. The instruments' greatest coup? Capturing the fossil whisper of creation, the remnant radiation from the Big Bang, now cooled down to a uniform wash of microwaves that blankets the universe.

Jansky, alas, saw none of this happen. Long burdened with a chronic kidney ailment, he died in 1950 at the early age of forty-four. In his last experiments, he was trying out a new-fangled gadget called a transistor to improve a radio amplifier.

Yet his legacy lives on with the new and improved Jansky Array in New Mexico, whose resolution and sensitivity are billions of times greater than those of the original merry-go-round. Even when the array is inevitably replaced or supplanted in the far future, Karl Jansky's name will still reverberate within the halls of radio astronomy. In 1973, the International Astronomical Union gave his name to a scientific unit. The *jansky* is a measure of the strength of an astronomical radio source.

CHAPTER TWENTY-THREE

The Once and Future Quasar
Discovering the universe's more violent side

I smiled when I heard the news. In 2017 an international team of astronomers had just announced the discovery of the most distant quasar, the luminous core of a newly forming galaxy situated a whopping 13 billion light-years away. That means the light from this quasar started on its journey less than a billion years after the Big Bang. The universe was just a toddler at the time.

I was amused because this headline has regularly been appearing in the news for more than half a century—and continues to this day. There's no news like old news. The most-distant-quasar record has gotten replaced as often as a newborn's diapers. It all started when Caltech astronomer Maarten Schmidt recognized the first quasar on February 5, 1963. And in doing so, he revealed an entirely new side to the universe's personality, one that both surprised and amazed astronomers. That's because they had all grown up thinking of the universe as fairly serene.

Hints that the early cosmos was edgier than once imagined had started arriving in the late 1950s. At that time the noted

British radio astronomer Martin Ryle reported that he counted more far-off cosmic radio sources than expected; the intense radio signals suggested that distant (and therefore, from our viewpoint, young) galaxies were more active than the older galaxies in our present-day universe. Spurred on by such discoveries in radio astronomy and not wanting to miss the boat, the United States built its own state-of-the-art radio observatories. One of them was a complex situated in California's Owens Valley and run by Caltech. Soon this observatory began studying radio sources with better resolution. So much so that in 1960 it was able to narrow down the location of a particularly strong source, labeled 3C 48 for being the forty-eighth object in the Third Cambridge Catalogue of radio sources.

Given these better coordinates, astronomer Allan Sandage then swiftly used the grand 200-inch (5-meter) Hale telescope atop southern California's Palomar Mountain to see what visible celestial object might be situated at that spot. Expecting to see a galaxy, he instead found a blue pinpoint of light, a real surprise. At first, everyone just assumed it was a star in our own galaxy, making it the first known "radio star." But there was a catch: "I took a spectrum the next night," said Sandage, "and it was the weirdest spectrum I'd ever seen."

Over the next two years, a handful of similar objects were discovered, adding to the mystery. On first look they appeared to be simply faint blue stars within the Milky Way, just like 3C 48. But again, the light waves emanating from these so-called radio stars displayed spectral features unlike those of any star ever observed. It was like riding down a familiar turnpike and finding all the road signs written in gibberish. Optical astronomers couldn't even find evidence that hydrogen, the main com-

Maarten Schmidt in the
1960s, when he recognized
the true nature of the first
quasar.
(AIP Emilio Segrè Visual Archives,
John Irwin Slide Collection)

ponent of all stars, was present in these objects. Yet, everyone kept assuming they were stars because, well, they *looked* like stars through an optical telescope. Not until February 1963 was the identity of these peculiar radio beacons finally unmasked.

On the fifth day of that month, the thirty-three-year-old Schmidt, who had arrived a few years earlier at Caltech from the Netherlands, was sitting at his desk attempting to write an article for the British journal *Nature* on the radio star known as 3C 273. He had just obtained an optical spectrum of this strange object, using the Hale telescope. With the spectrum spread before him, Schmidt came to recognize a familiar pattern of spectral lines that had eluded him for weeks. The pattern resembled the light waves typically emitted by simple hydrogen—but they were in the wrong place. That's why hydrogen had appeared to be missing! The hydrogen lines were there, but shifted *waaaay* over, toward the red end of the spectrum. That meant this starlike object was moving away from us at a tremendous speed. Just as the pitch of an ambulance siren gets lower as it races away, a light wave is stretched when its source recedes from us, and, because a light wave at the red end of the spectrum is longer, we say it gets "redder." This "redshift" lets astronomers gauge not only how fast a celestial object is moving but also its distance, because—as Edwin Hubble found in 1929—there's a systematic link between a galaxy's speed and its distance in our expanding universe. The faster the velocity, the more distant the galaxy.

In this way, Schmidt swiftly grasped what that redshift meant. 3C 273 was not an unusual star situated within the Milky Way, but rather a bizarre object located about two billion light-years away (one of the farthest cosmic distances ever recorded at that time). 3C 273 was rushing away from us at some 30,000 miles (48,000 kilometers) per second, carried outward with the swift expansion of the universe. Schmidt knew that only an incredibly bright source could be visible from such a distance; he figured 3C 273 was radiating the power of trillions of stars and suspected it was the brilliant and very disturbed nucleus of a distant galaxy. This galaxy appeared starlike only because it was so far away.

With that revelation, all fell into place. The spectra of other mystifying radio stars were quickly deciphered. These

An image of 3C 273, the first identified quasar,
taken by the Hubble Space Telescope.
(ESA/Hubble & NASA)

blue, extragalactic specks were soon christened quasi-stellar
radio sources (QSRS). Before long, they were simply called
quasars. For his role in vastly extending the boundaries of the
visible universe, Schmidt made the cover of *Time* magazine.

3C 273 is now considered relatively close to us, as quasars
go. Its distance is small potatoes compared with those of later
finds. Today's record holders are more than six times as far away.
And the fact that earthbound observers are able to photograph
such quasars across the vastness of the universe means that
these objects are the most powerful denizens of the heavens.

What could possibly be the source of a quasar's monstrous energy? That's the first thing everyone asked when 3C

273's secret was revealed. "The insult was not that they radiate so much energy," said Schmidt, "but that this energy was coming from a region probably no more than a light-week across." Astronomers came to know this by seeing the quasars dim and brighten over a matter of weeks or days. In the case of 3C 273, they checked old photographic plates of the 13th-magnitude object, going back some seventy years. In one picture it was faint, a month later it was brighter. Such relatively swift and sizable fluctuations meant that the quasar's power source was small, perhaps less than the diameter of our solar system. (Any small luminosity change in a vastly larger object would get lost in the noise.) Yet from such a cosmically tiny region spewed the energy of billions of suns. Tapping into such a cosmic dynamo for just one second would power the world for a billion billion years.

Since Schmidt's discovery, quasars have been closely examined by an array of telescopes—radio, infrared, optical, and X-ray. And all point to one answer to a quasar's identity: it's a supermassive black hole residing in the center of a young, gas-filled galaxy. The vast energies are likely released as matter spirals in toward the black hole, and also by the spinning hole itself acting as a powerful dynamo, causing huge beams of energy to shoot out of the black hole's north and south poles.

The center of our home galaxy, the Milky Way, was probably a quasar in the distant past. The black hole lurking there, estimated to contain the mass of around four million suns, is now fairly quiet, having grabbed all the nearby "food" it can get. Its engine is on idle, but this behemoth might wake up one day, perhaps as we slowly collide with our close neighbor, the Andromeda galaxy, about four billion years from now.

To the Big Bang and Beyond

The universe was completely refashioned over the past century. Shortly after a proper cosmic yardstick was devised and we learned that the Milky Way was accompanied by billions of other galaxies in the universe, we were further astounded to find out that space-time was expanding, with galaxies surfing outward on the wave. And it wasn't just Edwin Hubble who provided the evidence for this expansive behavior; he was helped by a former farm boy from the Midwest who is little known today, along with a Belgian priest.

Once the cosmic expansion was accepted, it didn't take long for scientists to imagine that ballooning in reverse, leading to the conception of the Big Bang. How to prove that our universe began with a mighty explosion, however, took time. What to look for was first revealed in 1948, but that prediction was not firmly proven for nearly two decades.

Since then, cosmologists have added new details to the story of our cosmic creation. For example, our universe may have begun with a brief moment of superaccelerated expansion, called inflation. The "bang" came at the end, when inflation's latent energy transformed into all the particles and radiation that surround us today. More than that, it's possible that parallel universes were generated in a similar way, meaning we reside within a "multiverse," side by side with other universes. But to prove that, theorists must first wrestle with the nature of time.

CHAPTER TWENTY-FOUR

Finding a Cosmic Yardstick

Henrietta Swan Leavitt's painstaking observations
inspired a new way to determine the distances to
far-off galaxies

IRST-TIME travelers to the Southern Hemisphere
might mistake the deep-space nebular clouds visible there for high cirrus formations, somehow
made luminous in the dark of night. Yet the Large
and Small Magellanic Clouds are each a chaotic collection of
stars, richly diffused with glowing gas. Such novel and fascinating sights were a compelling reason for early European
and American astronomers to set up observatories in the
Southern Hemisphere.

In the early 1890s, the Harvard College Observatory
established a southern station in the highlands of Peru. For
more than a decade, Harvard had been cataloging every star
in the northern sky and accurately gauging its color and
brightness. With a sizable endowment for a program in
spectroscopy, observatory director Edward C. Pickering resolved to further classify the brightest stars by their chemical
spectra. The Peruvian observatory allowed Harvard to extend

all those endeavors to the southern sky. In doing this, Pickering was helping astronomy move beyond just tracking the motions of stars across the sky to figuring out their basic properties.

With a huge number of glass photographic plates of the northern and southern skies stacking up, Pickering shrewdly recognized the value of smart young women yearning to contribute in an era that generally denied them full access to scientific institutions. These woman "computers," as they were called, some with college degrees in science, could be hired for less than half the pay of a man. Stationed at the observatory's headquarters in Cambridge, Massachusetts, they peered at plates all day through magnifying glasses, swiftly and accurately numbering each star, determining its exact position,

Henrietta Leavitt working at the Harvard College Observatory.
(AIP Emilio Segrè Visual Archives, *Physics Today* Collection)

and assigning it either a spectral class or a photographic magnitude.

One of Pickering's most brilliant hires was Henrietta Swan Leavitt, who began work as a volunteer soon after graduating, in 1892, from what later became Radcliffe College. She proved herself an expert in stellar photometry, gauging the magnitude of a star by assessing the size of the spot it imprinted upon a photographic plate. As she worked, she was also instructed to keep an eye out for variable stars, those that regularly increase and decrease in brightness.

Leavitt left Harvard for a time in 1896, first traveling through Europe for two years and then moving to Wisconsin to be with her father. But by 1902, she returned to Harvard as a paid employee, and within two years, variable stars came back into her life in full force.

Looking through a magnifying eyepiece at two plates of the Small Magellanic Cloud, taken at different times, she noticed that several stars had changed in brightness, as if they were undergoing a slow-motion twinkle. Over the following year, she looked at additional images of the cloud and found dozens more variable stars. Soon in her tally she included old plates going back to 1893, and then started examining the Large Magellanic Cloud as well. By 1907 she had found a record-setting total of 1,777 new variable stars within these prominent, mistlike clouds.

Leavitt dutifully reported her findings in the 1908 *Annals of the Astronomical Observatory of Harvard College*, paying particular attention to a special group of sixteen variable stars in the Small Magellanic Cloud. They were later identified as Cepheid variables, stars thousands of times more luminous

The Small and Large Magellanic Clouds (top left, bottom left)
as seen from Cerro-Tololo Inter-American Observatory in Chile.
The Milky Way is on the right.
(Roger Smith/NOAO/AURA/NSF/WIYN)

than our Sun. One sentence in Leavitt's report would become
her most venerated statement. "It is worthy of notice," she
wrote, "that . . . the brighter variables have the longer periods." Because all her Cepheids were situated in the Small
Magellanic Cloud, Leavitt could assume they were all roughly the same distance from Earth. Their periods, therefore,
were directly associated not only with their apparent brightness as seen from Earth, but with the actual emission of light.

Leavitt's discovery would lead to a new cosmic yardstick, one that would allow later astronomers to determine the distances to far-off celestial objects, which had never been measurable before.

Leavitt was on track to discover the celestial equivalents of lighthouses on Earth. A sailor at sea who knows the intensity of light emitted by a lighthouse can estimate how far away it is by how bright the beacon appears. Similarly, if an astronomer could know the absolute brightness of a Cepheid—how luminous it would appear up close—he could estimate how far away it must be to appear as the faint point of light seen from Earth. But, just as some lighthouses shine with brighter lights than others, so do Cepheids. Only their relative intensities can be measured from afar. The promise of Leavitt's discovery was this: if the absolute brightness of just one Cepheid could be known, the absolute brightness of the others could be figured out based on the differences in their periods. In this way, each Cepheid could become an invaluable "standard candle" (as astronomers call it) for gauging distances deep into space.

In 1908, however, Leavitt was wary that her initial sample of sixteen Cepheids was too small to secure a firm and predictable "period-luminosity" law. She needed more, but chronic illnesses, one of which had earlier left her deaf, and the death of her father delayed her a few years. Moreover, Cepheids, though very bright, are also very rare. Not until 1912 was Leavitt able to add nine more Small Magellanic Cepheids to her list. With twenty-five in hand, all at roughly the same distance from Earth, she could at last establish a distinct mathematical relationship between the rate of a Cepheid's blinking and its perceived brightness. In a logarithmic-scale graph of her data, the visible brightness of her Cepheids rises in a sure, straight diagonal line as the stars' periods get longer and longer. She had found her law.

Cepheids stood ready to be the perfect standard candles, but first Leavitt needed to know the true brightness of at least one. From that one, her graph could be calibrated such that an astronomer could pick out a far-off Cepheid anywhere in the sky, measure its period, and infer its actual luminosity. Knowing that, the star's distance could be calculated from its much fainter apparent brightness. First, however, Leavitt required the reverse: knowing the distance to one bona fide Cepheid was the only way to calculate its true brightness!

But Leavitt's going to a telescope to pursue an answer was out of the question, not only because women were denied access to the best telescopes at the time, but because of her frail condition. She had been advised by her doctor to avoid the chilly night air habitually braved by observers. If she had the know-how, she could have carried out a calculation from her desk, using stellar data from previously published work, but Pickering held the strong conviction that his observatory's prime function was to collect and classify data, rather than apply it to solve problems. He had other things for her to do. At his behest, Leavitt dedicated herself for several years to a separate project on stellar magnitudes. Ultimately, her work served as the basis for an internationally accepted system that is still in use, though now revised.

In the meantime, recognizing the value in Leavitt's truncated research, the Danish astronomer Ejnar Hertzsprung picked up where she left off. In 1913, he devised a statistical model using known Cepheids in the Milky Way to calibrate

Leavitt's period-luminosity graph. From that, he calculated the first intergalactic distance, to the Small Magellanic Cloud, thereby fulfilling the momentous promise of her work.

Yet Leavitt's desire to pursue further research on the variables never left her. Soon after Pickering's death in 1919, she at last divulged her interest to the observatory's soon-to-be director, Harlow Shapley. But just as she was on the verge of completing her prolonged stellar-magnitude project—when she might have at last returned to her work on variables—Henrietta Leavitt passed away, at the age of fifty-three. She had endured a grueling struggle with stomach cancer. By the time of her death, on December 12, 1921, she had discovered some 2,400 variable stars, about half the number then known to exist.

Unaware of Leavitt's passing, a member of the Royal Swedish Academy of Sciences four years after her death contacted the Harvard College Observatory to inquire about her discovery, intending to use the information to nominate her for a Nobel Prize in Physics. By the rules of the award, however, the names of deceased individuals could not be submitted.

Leavitt's work certainly deserved the prize. By the time the Swede's message had reached the Harvard College Observatory, her period-luminosity law had led to two momentous astronomical discoveries. It allowed Shapley in 1918 to demonstrate that our Milky Way was far larger than originally thought, with the Sun relocated away from the galactic center. And by 1923, Edwin Hubble spotted a Cepheid in the Andromeda nebula, which turned out to reside far beyond the borders of our galaxy. Leavitt's law helped prove that the Milky Way is not alone in the universe but just one of many galaxies.

CHAPTER TWENTY-FIVE

The Cosmologist Left Behind

Edwin Hubble usually gets the credit, but
Vesto Slipher was the first to see the signs
that the universe is expanding

A T the end of the nineteenth century, the wealthy
Bostonian Percival Lowell—the black sheep of
one of New England's leading families—built a
private observatory atop a pine-forested mesa in
Flagstaff, Arizona, to study Mars, its supposed canals, and its
presumed inhabitants. There, some 1.4 miles above sea
level, Lowell installed a 24-inch (61-centimeter) Alvan Clark
refractor—not a very large telescope even for the time, but
one perched higher than the 36-inch (91-centimeter) refractor at the venerable Lick Observatory in California.

This pleased Lowell immensely, for he sought to
outdo his California competitor at every turn. In 1900 he
ordered a custom-built spectrograph that was an improved
version of the one at Lick. To operate this new instrument,
Lowell hired a recent graduate of the Indiana University astronomy program: an Indiana farm boy named Vesto Melvin
Slipher.

Lowell chose well. Slipher took a spectrograph intended for planetary work and with great skill eventually extended the observatory's work far beyond the solar system. Instead of discerning new features on the Red Planet, the observatory's raison d'être, Slipher found himself confronting a surprising aspect of the wider cosmos, previously unknown. He detected the very first hint—the earliest glimmers of data—that the universe is expanding. But it took more than a decade for astronomers to fully recognize what he had done.

A century ago, when one-third of Americans lived on rural farms lit by only candle or kerosene, the nighttime sky

Vesto Slipher.
(From the Lowell Observatory Archives)

was breathtaking. The Milky Way arched across the celestial sphere like an army of ghosts. This sublime stellar landscape must have been a powerful inspiration, for many of America's greatest astronomers a century ago were born on Midwest farms, like Slipher.

"V. M.," as he was known to friends, must have had qualms upon arriving at Flagstaff in the summer of 1901. The biggest telescope he had ever operated was a 4.5-inch (11-centimeter) reflector. The young man struggled for a year to handle the spectrograph with ease. He even initially confused the red and blue ends of the spectrum on its black-and-white photographic plates, a scientific faux pas of the first magnitude. In distress, Slipher asked Lowell if he could go to Lick for instruction, but his boss firmly said no. Given the rivalry between the two observatories, Lowell didn't want Lick knowing that one of his staff needed help.

Slipher and Lowell were an intriguing mesh of personalities, like a harmony created from two different notes. Flamboyant and aggressive, Lowell hated to share the spotlight. Slipher was, fortunately, Lowell's opposite in character. A modest and reserved man, he knew it wasn't wise to steal Lowell's thunder. More than that, he didn't want to.

Slipher made progress on the spectrograph, eventually becoming a virtuoso at its operation. By 1909 he was able to confirm that thin gas existed in the seemingly empty space between the stars; it left spectral lines in starlight that were narrower and at slightly different Doppler shifts than the spectral lines arising in the stars' atmospheres. This triumph won him praise from astronomers around the world. In 1912 he determined that the faint Merope Nebula in the Pleiades had the same spectrum as the Pleiades stars themselves, the first proof of a reflection nebula made of interstellar dust ("pulverulent matter," he called it). In due course these pursuits led Slipher to his greatest discovery of all.

It began innocently enough. On February 8, 1909, Lowell in Boston sent a typed letter to Slipher with concise instructions: "Dear Mr. Slipher, I would like to have you take with your red sensitive plates the spectrum of a white nebula—preferably one that has marked centres of condensation." By "white nebula" Lowell meant what we now call a spiral galaxy. At the time, however, many astronomers assumed that these spiraling nebulae were nearby planetary systems under construction. Lowell stressed that he wanted "its outer parts." He longed to see if the chemical elements at a spiral nebula's edge, as revealed by their spectral lines, matched the composition of the giant planets in our outer solar system. A connection would mean the spirals could indeed be the precursors of planetary systems.

Slipher balked at first. "I do not see much hope of our getting the spectrum," he told Lowell. Photographic emulsions in 1909 had extremely slow speeds. Slipher knew that it would take at least a thirty-hour exposure to take just an ordinary photograph of the nebula with the long-focus refractor. To acquire a spectrum—what with light being lost in the spectrograph and the remaining light being spread out into a strip—seemed impossible.

Although Slipher considered the task hopeless, he persevered and by December 1910 was able to wrench some feeble data from the Great Nebula in Andromeda (M31). "This plate of mine," he informed Lowell by letter, "seems to me to show faintly peculiarities not commented upon." He was convinced that he had captured something on the spectrum previously unseen by other spectroscopists.

By trial and error, Slipher made improvements to the spectrograph. Instead of using a set of three prisms, which separated spectral lines widely, he decided to use just one, which reduced the light loss and also spread out the light less on the plate. He also understood that increasing the speed of the system was vital; he bought a very fast, commercially available photographic lens to go ahead of the plate.

Planet studies and reports on the return of Halley's Comet kept Slipher from getting back to the Andromeda Nebula until the fall of 1912. But by then his refashioned spectrograph was operating two hundred times faster than its original specifications, allowing him to slash his long exposure times. He could at last try for the spectrum he had so long sought.

Slipher made his first exposure with the new system on September 17. It took a total of six hours and fifty minutes for Andromeda's faint light to fully register. "It is not really very good and I am of the opinion that we can do much better," he relayed to Lowell. He soon acquired two more spectra. When carrying out these observations, the interior of the wooden dome at times could resemble the movie version of a mad scientist's laboratory, with a high-voltage induction coil sparking and sputtering by the side of the telescope. A row of old-fashioned Leyden jars served as capacitors to juice up the sparks. This contraption served to vaporize traces of iron and vanadium inside the spectrograph; the light of the sparks passed through the spectrograph and onto the photographic plate. The known emission spectra of the vaporized elements provided the calibration lines needed to measure the exact wavelengths of the absorption lines in the nebula's spectrum.

Each spectrum that Slipher produced was tiny: just a centimeter long and a millimeter wide. He needed a microscope to measure how much the spectral lines might have been Doppler-shifted from their rest wavelengths. The microscope was with Lowell in Boston temporarily, and Slipher didn't get it back until mid-December. But once it arrived, he couldn't resist taking a quick peek at the Andromeda plates he had so far. There were "encouraging results or (I should say) indications," Slipher reported to Lowell, "as there appears to be an appreciable displacement of the nebular lines toward the violet." A shift of the lines toward the blue-violet end of the spectrum meant that Andromeda was moving *toward* Earth.

But Slipher felt he needed a better spectrum to measure the speed accurately. He started the final exposure on December 29 and stayed with it until clouds rolled in near midnight. On a seeing-quality scale from 1 to 10—1 being the worst, 10 the best—Lowell astronomers often joked that at 10 you can see the Moon, at 5 you can still see the telescope, and at 1 you can only feel the telescope. Fortunately, the sky was clear the following night, and he was able to collect additional light for nearly seven more hours. Perhaps pressing his luck, he went into a third night, New Year's Eve.

Throughout January 1913 Slipher focused on measuring his plates with utmost precision. The result astonished him. The Andromeda Nebula was rushing toward the solar system at the ridiculous speed of 300 kilometers (186 miles) per second (a total of 670,000 miles per hour). This was about ten times faster than the average motion of stars in the Milky

Way. If the nebula was really a nearby star and planetary system in formation, it was wildly abnormal.

Instead of announcing this result in a major astronomical journal, Slipher chose to publish a brief account in the *Lowell Observatory Bulletin*. True to form, Slipher held off any grander statement until he had some confirmation. Yet even one spiral-nebula velocity was an exceptional accomplishment. Lowell was enormously pleased. "It looks as if you had made a great discovery," he wrote Slipher. "Try some more spiral nebulae for confirmation."

Working on Andromeda, though, was a holiday compared with gathering enough light from other white nebulae. Andromeda is the biggest spiral in the sky; the others only get smaller and dimmer, which made it even harder for Slipher to obtain their velocities. "Spectrograms of spiral nebulae are becoming more laborious now because the additional objects observed are increasingly more faint and require extremely long exposures that are often difficult to arrange and carry through owing to Moon, clouds and pressing demands on the instrument for other work," he noted.

Slipher's first target after Andromeda was M81. He then worked on a peculiar nebula in southern Virgo, NGC 4594, which he described as a "telescopic object of great beauty." It's now known as the Sombrero galaxy. Slipher eventually found that it was moving at a speed "no less than three times that of the great Andromeda Nebula." This time, however, the nebula was not traveling toward us, but *away*—at some one thousand kilometers (620 miles) per second. Slipher was greatly relieved. Finding a nebula that was racing outward rather than

A Hubble Telescope view of NGC 4594, known as the Sombrero galaxy. In 1913, Vesto Slipher measured this object as moving away from the Milky Way at some 1,000 kilometers per second.
(NASA, ESA, Hubble Heritage Team [STSci/AURA])

approaching removed any lingering doubts that the velocities might not be real. "When I got the velocity of the Andr. N. I went slow for fear it might be some unheard-of physical phenomenon," he wrote his former Indiana professor John Miller.

In the succeeding months Slipher kept expanding his list. His accomplishment was all the more amazing considering the relative crudeness of his instrument. The 24-inch telescope had only manual controls, and they weren't yet sophisticated enough for fine guiding. Yet Slipher had to hold the tiny image of each nebula on the slit of the spectrograph steadily for hours on end as the telescope tracked the turning sky. When asked years later how he was able to do this, Slipher replied dryly, "I leaned against it."

By the summer of 1914 Slipher had the velocities of fourteen spiral nebulae in hand. And with this collection of data, an undeniable trend at last emerged: While a few nebulae, such as Andromeda, were approaching us, the majority were rapidly moving away.

Suddenly the older idea that the white nebulae were other galaxies—other "island universes" of stars at fantastically great distances (an idea dating from Immanuel Kant in 1755)—looked newly plausible. "It seems to me, that with this discovery the great question, if the spirals belong to the system of the milky way or not, is answered with great certainty to the end, that they do not," Danish astronomer Ejnar Hertzsprung wrote Slipher. The speeds were too great for them even to stay within our home galaxy. But Slipher at this stage was still on the fence: "It is a question in my mind to what extent the spirals are distant galaxies," he responded. But he was absolutely sure of his velocity measurements.

For most of his career Slipher published few detailed papers of his work outside of Lowell's in-house bulletin. He published very little at all from 1933 until his retirement in 1954, having turned much of his attention to local business pursuits and community affairs. The great standout in his otherwise sparse research record was his work on spiral-nebula velocities. He was absolutely confident of what he was seeing—so confident that he for once overcame his homebound nature and traveled in August 1914 to Northwestern University in Evanston, Illinois, to present his results in person.

At Northwestern, sixty-six astronomers from around the United States gathered by Lake Michigan for their annual meeting. Slipher reported in his talk that the average speed of the spirals was now "about 25 times the average stellar velocity." Of the fifteen spiral nebulae he had measured so far, three were approaching Earth and the rest were moving away. The velocities ranged from "small," as it was recorded on his list, to an astounding 1,100 kilometers (680 miles) per second, the greatest speed of a celestial object ever measured up to then. When Slipher finished delivering this remarkable news, his fellow astronomers rose to their feet and gave him a resounding ovation. No one had ever seen such a spectacle at an astronomical meeting. And with good reason: Slipher alone had climbed to the top of the Mount Everest of spectroscopy. In the audience was a young, ambitious astronomer named Edwin P. Hubble, just starting his graduate degree, who would later seize on Slipher's work and extend it.

After a few more years, the cautious Slipher at last came around to Hertzsprung's view and began to envision the Milky Way as moving among other galaxies just like itself. He even speculated in 1917 that the spirals might be "scattering" in some way—a precocious intimation of cosmic expansion that took many more years to fully recognize. But acceptance of spiral nebulae as distant galaxies could not be fully achieved until astronomers could determine how far away Andromeda and its sister nebulae truly were.

That, of course, famously occurred in 1923–24 when Hubble, using the 100-inch telescope on California's Mount Wilson, identified Cepheid variable stars within Andromeda and used their pulsation periods as cosmic yardsticks to establish that the nebula was indeed a separate island universe. Five years later, in 1929, working with Milton Humason, Hubble identified a mathematical trend in the flight of the galaxies. The velocity at which the galaxies were moving away from us steadily increased as he peered ever deeper into space. The greater the distance of the nebula, the higher its velocity. The numerical value describing this trend became known as the "Hubble constant."

Hubble was quite possessive of this finding and kept close watch on it. When Dutch astronomer Willem de Sitter, in a 1930 review article, casually referred to several astronomers linking a galaxy's velocity to its distance, Hubble picked up his pen and reminded de Sitter who should receive the lion's share of the credit. "I consider the velocity-distance relation, its formulation, testing and confirmation, as a Mount Wilson contribution and I am deeply concerned in its recognition as such," he wrote.

Hubble conveniently forgot to tell de Sitter that the galaxy velocities he first drew upon in his historic 1929 paper were actually Slipher's old data, which Hubble used without acknowledgment, a serious breach of scientific protocol. Hubble partially made up for this nefarious deed much later, in 1953. As Hubble was preparing a talk, he wrote Slipher, asking for some slides of his first 1912 spectrum of the Andromeda Nebula, and in this letter he at last gave the Lowell Observatory astronomer due credit for his initial breakthrough. "I regard such first steps as by far the most important of all," wrote Hubble. "Once the field is opened, others can follow." In the lecture itself, Hubble professed that his discovery "emerged from a combination of radial velocities measured by Slipher at Flagstaff with distances derived at Mount Wilson." Privately, Slipher was bitter that he didn't receive more immediate public credit but was too humble to demand his share of the glory.

In some ways, Slipher's accomplishment resembled that of Arno Penzias and Robert Wilson several decades later. In 1964 the two Bell Laboratories researchers were calibrating a horn-shaped antenna in New Jersey in preparation for some radio observations and found unexpected static wherever they pointed. Just as Slipher made a remarkable cosmological find that took others time to fully interpret, Penzias and Wilson needed fellow astronomers to tell them what they had found: the afterglow of the Big Bang. But whereas Penzias and Wilson received the Nobel Prize for their serendipitous discovery, Slipher, as the years passed, was nearly forgotten in the momentous saga of the fleeing galaxies. A namesake like the "Slipher Space Telescope" was never to be.

CHAPTER TWENTY-SIX

The Primeval Atom

A Belgian cleric laid the groundwork for both the
expanding universe and the Big Bang

THE idea that the universe is expanding was one
of the most revolutionary and unsettling findings of modern astronomy. As seen in the previous chapter, the germ of the idea arose not solely with Edwin Hubble at the Mount Wilson Observatory in
California in 1929, as so many textbooks suggest. In addition,
we must look to the halls of MIT and Harvard a few years
before Hubble even initiated his historic measurements of
galaxy distances and motions. There the very theory of an
expansion was hatched in the mind of a Jesuit priest, who was
studying at MIT's physics department.

A military hero, Georges Lemaître had received the Croix
de Guerre for his service in the Belgian artillery after Germany invaded his homeland in World War I. He went on to
earn a doctorate in mathematics at the Catholic University
of Louvain; afterward, perhaps affected by the horrors he
had observed from the trenches, he enrolled in a seminary.
Although he was ordained in 1923, the Church permitted him

to continue his scientific pursuits. Captivated by the beauty of Einstein's new general theory of relativity, the abbé proceeded to the University of Cambridge to broaden his understanding of the theory's equations under the guidance of the astrophysicist Arthur Eddington, who deemed his student "exceptionally brilliant."

In 1924, after a year in England, Lemaître traveled to the United States to study at Harvard's observatory and enroll in MIT's Ph.D. program in physics. His dark hair combed straight back and his cherubic face adorned with round glasses, he could easily be spotted on the college campuses by his attire—a black suit or an ankle-length cassock, set off by a stiff white clerical collar. Some could find him just by following the sound of his full, loud laugh, which was readily aroused.

In pursuit of his second Ph.D., Lemaître became interested in applying general relativity to the universe at large, which many in the 1920s believed to consist entirely of our own galaxy. By then totally absorbed by astronomy, he made sure to attend the 1925 meeting of the American Astronomical Society in Washington, D.C., where a crucial discovery was announced: Edwin Hubble had proved that certain spiral nebulae, previously thought to be gaseous clouds within the Milky Way, were actually separate galaxies far beyond its borders.

While others in the room were focused on Hubble's revelations about the true nature of these long-perplexing nebulae, Lemaître was two jumps ahead. Though new to astronomy, he quickly realized that the newfound galaxies could be used to test certain predictions that general relativity made about the universe's behavior. Soon after the meeting, Lemaître began formulating his own cosmological model.

Georges Lemaître (left) and Albert Einstein in
1933 at the California Institute of Technology.
(Courtesy of the Archives, California Institute of
Technology)

Two models were already in circulation in the astrophysical community. According to the first, proposed by Einstein himself in 1917, the universe contained so much matter that space-time wrapped itself up into a hyperdimensional ball—a closed, stable, enduring system. The second, posited soon after by the Dutch astronomer Willem de Sitter, was very different: it assumed that cosmic densities were so low that the universe could be considered empty. The unique properties of space-time that arose in this model caused light waves to get longer the farther they traveled from their source. This aspect of the model was consistent with some recent astronomical news that de Sitter was well aware of, but Einstein wasn't. At the Lowell Observatory in Arizona, most of Vesto Slipher's spiral nebulae spectra were shifted to the red, a shift that implied the nebulae were moving outward into space— indeed, at the greatest celestial velocities that had ever been observed (as noted in the previous chapter). But de Sitter posited that the nebulae might only *appear* to be moving; instead, he suggested, the light waves themselves were getting longer and longer as the light traveled toward Earth.

Lemaître was not comfortable with either model. De Sitter's could explain the redshifted nebulae but required a universe that was empty (which he was sure it was not); Einstein's accommodated a universe filled with matter but couldn't account for the fleeing nebulae. Lemaître aimed, as he put it, to "combine the advantages of both."

While studying at MIT, Lemaître visited Slipher at the Lowell Observatory and Hubble at Mount Wilson to learn the latest velocity and distance measurements for what were now known to be spiral galaxies. With this information in hand, he took a first stab at a new solution, but he had not fully developed it by the end of 1925, when he handed in his Ph.D. thesis and left MIT. His thesis contained a preliminary model, a modification of de Sitter's view of the universe. On returning to Belgium, where he became a professor at the Catholic University of Louvain, he fleshed out that modification into an entirely new model, which he published in 1927. Nearly two full years before Hubble provided the definitive observational proof, Lemaître unveiled a cosmological model in which space-time continually stretches, and galaxies move outward on the wave. (The gravitational field of a galaxy, far stronger than the field outside it, keeps the galaxy intact during the expansion.) The galaxies' retreat, he wrote in his paper, is "a cosmical effect of the expansion of the universe." He even estimated a rate of expansion, a number close to the figure that Hubble eventually calculated and which came to be known as the "Hubble constant."

This was a tremendous accomplishment and offered an astounding vision of how the universe operates. But no one noticed—no one at all. Lemaître's paper was completely ignored, probably because he inexplicably published it in an obscure Belgian journal. A similar solution, conceived independently in 1922 by the Russian mathematician Aleksandr Friedmann, went unnoticed as well. At a 1927 meeting in Brussels, Lemaître cornered Einstein and tried to persuade him to accept this new vision of the universe. But the world-renowned physicist would have none of it. "Your calculations are correct, but your physical insight is abominable," he replied. Einstein refused to imagine a universe in which space-time was stretching.

This impasse stood for a couple of years. But in 1929, Hubble verified that the galaxies were moving outward in a uniform way. And in 1931 Lemaître's paper was finally noticed by Eddington and consequently reprinted in the more prominent *Monthly Notices of the Royal Astronomical Society*. Why Hubble saw the velocities of the galaxies steadily increase with distance was finally explained. Only then was the expanding universe truly recognized. Astronomers and theorists alike were thunderstruck by this radically new cosmic setup, breathtaking in its grandeur and terrifying in its implications.

Perhaps most consequential was the question that Lemaître first posed in his original 1927 paper: How did this expansion get started? "It remains to find the cause," he wrote at the time. But within four years he boldly suggested in the journal *Nature* that all the mass-energy of the universe was once packed within a "unique quantum," which he later called the primeval atom. From Lemaître's poetic scenario arose the current vision of the Big Bang, a model that shapes the thought of cosmologists today as strongly as the idea of crystalline spheres, popularized by Ptolemy, influenced natural philosophers in the Middle Ages.

Unlike Galileo, who was condemned to house arrest for his defense of a Sun-centered universe, Lemaître was lauded by the Church for his cosmic breakthrough. Indeed, he ultimately rose to the rank of monsignor and was made a fellow and later president of the Pontifical Academy of Sciences. But he recoiled from any suggestion that his primeval atom had been inspired by the biblical story of Genesis. Throughout his life, he insisted that his theory about the origin of space and time expanding outward from a quantum nugget sprang solely from the equations before him.

Lemaître made few notable contributions to cosmology after the 1930s, spending more time on celestial mechanics and pioneering the use of electronic computers for numerical calculations. But he continued to hope that the explosive origin of the universe would be validated by astronomical observations.

In June 1966, as Lemaître was fighting leukemia, Odon Godart, his successor at the Belgian university, visited him at the Hospital of Saint Peter with news of a report that had appeared in the *Astrophysical Journal* the previous year. That report, which would later win the Nobel Prize for Arno Penzias and Robert Wilson, had detailed the discovery of the cosmic microwave background; Godart brought confirmation that this was the remnant echo of the Big Bang. Lemaître died a few days later, on June 20, knowing that the universe was indeed launched from a compact bundle of energy, just as he had posited nearly four decades earlier.

CHAPTER TWENTY-SEVEN

Proving the Big Bang

Sometimes scientists don't realize the answer is
hidden in plain sight

S OMETIMES a great scientific idea needs time to take
root. Sometimes the world simply isn't ready. Con-
tinental drift comes to mind as an example, as well as
germ theory. Continents moving about? Microscopic bugs? Each of those propositions when first proposed
seemed too bizarre to accept right off. In such situations,
scientists have to be convinced that a new concept is worth
looking into.

Astronomy is no exception. A famous case is a prediction
of cosmic proportions that first appeared in a 1948 scientific
paper almost as an afterthought—and was soon forgotten.
Decades passed before the dismissed conjecture turned into
cosmology's greatest tool.

By the late 1940s, scientists had been grappling for
several years with a tough question: how did the universe
come to manufacture its vast array of elements? Until the end
of the nineteenth century, everyone had just assumed that
matter always was and would always be, but revelations

coming out of atomic physics laboratories in the first half of the twentieth century—ranging from radioactivity to nuclear transformations—overturned that notion. The elements obviously came from somewhere. The most plausible factory was inside a star, but no physicist in that era could get stellar models to build an atom heavier than helium. Anything more weighty quickly disintegrated within their theoretical computations.

What to do? In 1942 the Russian-American physicist George Gamow simply looked around for another locale for cooking up the elements, and he found one in Georges Lemaître's "primeval atom." The idea, a relatively new one, was that the universe had emerged and expanded from an initial hot plasma. (The term "Big Bang" didn't arrive until 1949.)

As Gamow's graduate student at George Washington University in the mid-1940s, Ralph Alpher took on the challenge for his doctoral thesis and demonstrated theoretically how it could be done. Like some skilled astrophysical chef, he started with a highly compressed stew of neutrons that Gamow had nicknamed "ylem," after Aristotle's name for the basic substance out of which all matter was supposedly derived. As the temperature of the cosmos began to plunge, some of those particles decayed into protons, which promptly began to stick to remaining neutrons. Step by step, each element was built up from the one before it—from helium to lithium, lithium to beryllium, beryllium to boron, and so on through the periodic table. In less than half an hour, when the last of the free neutrons decayed away, the cosmic meal was complete, with Alpher and Gamow concocting the full complement of universal "flavors," all the way up to uranium.

Their first report on this mathematical recipe, a one-page synopsis published in *Physical Review*, is more famous for its by-line than its content. Gamow, a merry prankster, listed the paper's authors as Alpher, Bethe, and Gamow, even though noted physicist Hans Bethe never participated in the work. Gamow couldn't resist the pun on the first three letters of the Greek alphabet: alpha, beta, gamma. That the 1948 paper chanced to be published on April Fool's Day only added to the fun.

While earning his master's and Ph.D., Alpher had also been working at the Applied Physics Laboratory of Johns Hopkins University. There, after getting his doctorate, he continued to collaborate on Gamow's campaign to study the physics of the Big Bang model. He was joined by fellow lab employee Robert Herman. The two young scientists went on to develop a detailed account of the evolution of the newborn universe, work described in 1977 by physicist Steven Weinberg in his book *The First Three Minutes* as "the first thoroughly modern analysis of the early history of the universe."

Early in their investigations, the pair came to realize that Alpher's original scheme for elemental cooking had an insurmountable flaw: while the newborn universe could make a few light elements, the cosmic expansion both dispersed and cooled the primordial plasma before the heavier elements had any chance of forming. With better stellar models, others would later prove that stars could do the job after all (see chapter 15). But no matter: in the course of their investigations, Alpher and Herman were still able to make a historic calculation that has stood the test of time.

This result was revealed in an unusual manner. On October 30, 1948, Gamow published an article in the British

In 1949 a composite picture was constructed with Robert Herman
on the left, Ralph Alpher on the right, and George Gamow in the
center as a genie coming out of a bottle of "ylem," the proposed
mixture of elementary particles out of which the elements formed.
(American Institute of Physics, Center for History of Physics)

journal *Nature* titled "The Evolution of the Universe." But in
checking over Gamow's reported results, Alpher and Herman
found some errors. They soon dashed off a correction, a brief
letter to the editor barely four paragraphs long that was published within two weeks. With their more accurate figures,
Alpher and Herman showed how the density of matter and
the density of radiation changed as the universe evolved. In

doing so, they curtly noted at the end of their letter that "the temperature in the universe at the present time is found to be about 5° Kelvin." That's only 5 Kelvin above absolute zero, the point at which all motion ceases. (On the Fahrenheit scale, that's 9 degrees above absolute zero, which is –459.67 F.)

With little fanfare, Alpher and Herman were telling the world that the present-day universe is bathed in a uniform wash of radiation left over from the flood of highly energetic photons released in the fury of the Big Bang. Cooled down over the eons with the expansion of the cosmos, the waning fire now surrounds us as centimeters-long radio waves. Today it is known as the cosmic microwave background radiation (CMBR).

When their note was published, the primeval atom theory was still highly controversial. Many astronomers preferred the steady-state model of the universe, a theory that postulated that space-time had neither a beginning nor an end. But Alpher and Herman's calculation was a clear-cut means of deciding between the two opposing theories of the universe's behavior.

Yet no one followed up. Looking back, it's hard to fathom why astronomers in the 1950s didn't jump at the chance to point their instruments at the sky and capture this primordial whisper of creation. But some thought radio telescopes weren't yet sensitive enough for the task; and when a few astronomers did peg an overall temperature of interstellar space at around 3 K, they didn't link it to cosmology at all. Some of them thought it was an error in their instruments.

Radio astronomers may have been unresponsive because their field was just establishing itself after World War II, and cosmological tests were not yet taken seriously. As Weinberg noted, they "did not know that they ought to try" to detect the background radiation. The radio sky was all so new. There were too many objects—radio stars, radio nebulae, radio galaxies—grabbing their attention. Amid such distractions, Alpher and Herman's prediction was either dismissed or utterly overlooked. And since both men later went into industrial research, the two didn't have the opportunity to keep pushing astronomers to take a look, although they did try—at one point even holding a press conference to generate attention, but to no avail.

The idea didn't resurface until the mid-1960s, when a team of astrophysicists at Princeton University (and some Soviet cosmologists independently) again reasoned that the Big Bang's residual heat must be permeating the universe. At the same time, two Bell Lab researchers in New Jersey, Arno Penzias and Robert Wilson, accidentally detected what proved to be the primeval microwaves. They were trying to eliminate excess noise in a horn antenna they were calibrating for astronomical work, but a stubborn residue always remained. Once Penzias and Wilson learned of the Princeton team's work, they at last understood that their radio interference was cosmic. In 1965 the two groups published papers simultaneously in the *Astrophysical Journal*. Neither paper mentioned Alpher and Herman's earlier contribution. For detecting the cosmic microwave background radiation, Penzias and Wilson received the 1978 Nobel Prize in Physics.

Herman died in 1997, Alpher ten years later. Both were deeply pained that the career rewards for making their momentous prediction never came to pass for them—such as election to prestigious academies, sizable research grants, prized promotions. The honors that were bestowed arrived late (Alpher received the National Medal of Science in 2007, when he was hospitalized with his final illness). "But we should not indulge in sermonizing about the nature of science," the two noted in a scientific memoir of their work published in 2001. "On to more about the CMBR," they proclaimed. And so it should be.

Over the past two decades, detectors in space have measured the cosmic microwave background, now pegged at 2.7 K, in exquisite detail. By mapping the barely perceptible ups and downs of this signal across the breadth of the celestial sky, astronomers have revealed a wealth of cosmological information. They've viewed the quantum jiggles that led to galaxy formation, tallied the exact amount of ordinary matter contained in the universe, verified that there is five times more cosmic stuff of an unknown nature (called dark matter), and confirmed that space-time is permeated with a dark energy that is causing the universe not just to steadily expand, but to accelerate outward like a runaway drag racer. And to think that all this knowledge was gleaned from a radio murmur, a faint heat first mentioned unceremoniously in a brief note tucked away in a scientific journal around seven decades ago.

CHAPTER TWENTY-EIGHT

It's Now Einstein's Universe

His theories explain the universe we observe today

O N January 29, 1931, the world's premier physicist, Albert Einstein, and its foremost astronomer, Edwin Hubble, settled into the plush leather seats of a sleek Pierce-Arrow touring car for a visit to Mount Wilson in southern California. They were chauffeured up the long, zigzagging dirt road to the observatory complex on the summit, nearly a mile above Pasadena. Home to the largest telescope of its day, Mount Wilson was the site of Hubble's astronomical triumphs. In 1923–24 he had used the telescope's then colossal 100-inch mirror to confirm that our galaxy is just one of countless "island universes" inhabiting the vastness of space. Five years later, after tracking the movements of these spiraling disks, Hubble and his assistant, Milton Humason, had confirmed something even more astounding: The universe is swiftly expanding, carrying the galaxies outward.

On the peak that bright day in January, the fifty-one-year-old Einstein delighted in the telescope's instruments.

Like a child at play, he scrambled about the framework, to the consternation of his hosts. Nearby was Einstein's wife, Elsa. Told that the giant reflector was used to determine the universe's shape, she reportedly replied, "Well, my husband does that on the back of an old envelope."

That wasn't just wifely pride. Years before Hubble verified cosmic expansion, Einstein had fashioned a theory, general relativity, that could explain it. In studies of the cosmos, it all goes back to Einstein.

Just about anywhere astronomers' observations take them—from the nearby Sun to the black holes in distant galaxies—they enter Einstein's realm, where time is relative, mass and energy are interchangeable, and space can stretch and warp. His footprints are deepest in cosmology, the study of the universe's history and fate. General relativity "describes how our universe was born, how it expands, and what its future will be," says Alan Dressler of the Carnegie Observatories. Beginning, middle, and end—"all are connected to this grand idea."

At the turn of the twentieth century, thirty years before Einstein and Hubble's rendezvous at Mount Wilson, physics was in turmoil. X-rays, electrons, and radioactivity were just being discovered, and physicists were realizing that their trusted laws of motion, dating back more than two hundred years to Isaac Newton, could not explain how these strange new particles flit through space. It took a rebel, a cocky kid who spurned rote learning and had an unshakable faith in his own abilities, to blaze a trail through this baffling new territory. This was not the iconic Einstein—the sockless, rumpled character with baggy sweater and fright-wig coiffure—but a

Albert Einstein with Edwin Hubble (behind Einstein, second from left) and others from Caltech and the observatory outside the dome of the 100-inch telescope during Einstein's visit to Mount Wilson on January 29, 1931.
(Courtesy of the Archives, California Institute of Technology)

younger, more romantic figure with alluring brown eyes and wavy hair. He was at the height of his prowess.

Among his gifts was a powerful physical instinct, almost a sixth sense for knowing how nature should work. Einstein thought in images, such as one that began haunting him as a teenager: If a man could keep pace with a beam of light, what would he see? Would he see the electromagnetic wave frozen in place like some glacial swell? "It does not seem that something like that can exist!" Einstein later recalled thinking.

He came to realize that since all the laws of physics remain the same whether you're at rest or in steady motion, the speed of light has to be constant as well. No one can catch up with a light beam. But if the speed of light is identical for all observers, something else has to give: absolute time and space. Einstein concluded that the cosmos has no universal clock or common reference frame. Space and time are "relative," flowing differently for each of us depending on our motion.

Einstein's special theory of relativity, published in 1905, also revealed that energy and mass are two sides of the same coin, forever linked in his famed equation $E = mc^2$. (E stands for energy, m for mass, and c for the speed of light.) "The idea is amusing and enticing," wrote Einstein, "but whether the Almighty is . . . leading me up the garden path—that I cannot know." He was too modest. The idea that mass could be transformed into pure energy later helped astronomers understand the enduring power of the Sun. It also gave birth to nuclear weapons.

But Einstein was not satisfied. Special relativity was just that—special. It could not fully describe all types of motion, such as objects in the grip of gravity, the large-scale force that shapes the universe. Ten years later, in 1915, Einstein made up for the omission with his general theory of relativity, which amended Newton's laws by redefining gravity.

General relativity revealed that space and time are linked in a flexible four-dimensional fabric that is bent and indented by matter. In this picture, Earth orbits the Sun because it is caught in the space-time hollow carved by the Sun's mass, much as a rolling marble would circle around a bowling ball sitting in a trampoline. The pull of gravity is just matter sliding along the curvatures of space-time.

Einstein shot to the pinnacle of celebrity in 1919, when British astronomers actually measured this warping. Monitoring a solar eclipse, they saw streams of starlight bending around the darkened Sun. With this new insight into gravity, physicists at last were able to make actual predictions about the universe's behavior, turning cosmology into a science.

Einstein was the first to try, an episode that showed that even he was a fallible genius. A misconception about the nature of the universe led him to propose a mysterious new gravitational effect (a notion he soon rejected.) But we now know he may have been right all along, and his "mistake" may yet turn out to be one of his deepest insights.

For Newton, space was eternally at rest, merely an inert stage on which objects moved. But with general relativity, the stage itself became an active player. The amount of matter within the universe sculpts its overall curvature. And his equations show that space-time itself can be either expanding or contracting.

When Einstein announced general relativity in 1915, he could have taken the next step and declared that the universe was in motion, more than a decade before Hubble directly measured cosmic expansion. But at the time, astronomers conceived of the universe as a large collection of stars fixed forever in the void. Einstein accepted this picture of an immutable cosmos. Truth be told, he liked it. Einstein was often leery of the most radical consequences of his ideas.

But because even a static universe would eventually collapse under its own gravity, he had to slip a fudge factor into the equations of general relativity—a cosmological constant. While gravity pulled celestial objects inward, this extra gravitational effect—a kind of antigravity—pushed them apart. It was just what was needed to keep the universe immobile, "as required by the fact of the small velocities of the stars," Einstein wrote in 1917.

Twelve years later, Hubble's verification that other galaxies were racing away from ours, their light waves stretched and reddened by the expansion of space-time, vanquished the static universe. It also eliminated any need for a cosmological constant to hold the galaxies steady. During his 1931 California visit, Einstein acknowledged as much. "The red shift of distant nebulae has smashed my old construction like a hammer blow," he declared. He reputedly told a colleague that the cosmological constant was his biggest blunder.

With or without that extra ingredient, the basic recipe for the expanding universe was Einstein's. But it was left to others to identify one revolutionary implication: a moment of cosmic creation. In 1931 the Belgian priest and astrophysicist Georges Lemaître put the fleeing galaxies into reverse and imagined them eons ago merged in a fireball of dazzling brilliance. "The evolution of the world can be compared to a display of fireworks that has just ended: some few red wisps, ashes and smoke," wrote Lemaître. From this poetic scenario arose to-day's Big Bang.

Many were appalled by this concept. "The notion of a beginning . . . is repugnant to me," said British astrophysicist Arthur Eddington in 1931. But evidence in its favor slowly gathered, climaxing in 1964, when scientists at Bell Telephone Laboratories discovered that the cosmos is awash in a sea of microwave radiation, the remnant glow of the universe's thunderous launch.

The high priests of astronomy have continued the cosmological quest initiated by Einstein and Hubble, first at Mount Wilson, then at the 200-inch telescope on California's Palomar Mountain, ninety miles (145 kilometers) to the south. How fast is the universe ballooning outward? they asked. How old is it? "Answering those questions," says Wendy Freedman, former director of the Carnegie Observatories, "turned out to be more difficult than anyone anticipated."

Only at the turn of this century, with the help of a space telescope aptly named Hubble, did Freedman and others confidently peg the universe's current rate of expansion, as well as its age. A birthday cake for the universe would require around 14 billion candles.

Astronomers have found some strange objects in this expanding universe—and these too are Einstein's children. In the 1930s a young Indian physicist, Subrahmanyan Chandrasekhar, applied special relativity and the new theory of quantum mechanics to a star. He warned that if it surpassed a certain mass, it would not settle down as a white dwarf at the end of its life (as our Sun will). Instead, gravity would squeeze it down much further, perhaps even to a singular point.

Chandrasekhar had opened the door for others to contemplate the existence of the most bizarre stars imaginable. First there was a naked sphere of neutrons just a dozen miles wide born in the throes of a supernova, the explosion of a massive star. Then there was the peculiar object formed from the collapse of an even bigger star or a cluster of stars— enough mass to dig a pit in space-time so deep nothing can ever climb out.

Einstein himself tried to prove that such an object—a black hole, it was later christened—could not exist. He loathed what would be found at a black hole's center: a point of zero volume and infinite density, where the laws of physics break down. The discoveries that might have forced him to acknowledge his theory's strange offspring came after his death in 1955.

Astronomers identified the first quasar, a remote young galaxy disgorging the energy of a trillion suns from its center, in 1963. Four years later, much closer to home, observers stumbled on the first pulsar, a rapidly spinning beacon emitting staccato radio beeps. Meanwhile, spaceborne sensors spotted powerful X-rays and gamma rays streaming from points around the sky. All these new, bewildering signals are believed to pinpoint collapsed objects—neutron stars and black holes—whose crushing gravity and dizzying spin turn them into dynamos. With their discovery, the once sedate universe took on an edge; it metamorphosed into an Einsteinian cosmos, filled with sources of titanic energies that can be understood only in the light of relativity.

Even Einstein's less celebrated ideas have had remarkable staying power. As early as 1912 he realized that a faraway star can act like a giant spyglass, its gravity deflecting passing light rays and magnifying objects behind it. (See chapter 17.) But he eventually concluded this effect had "little value." With to-day's telescopes, though, astronomers are seeing galaxies and galaxy clusters act as gravitational lenses, offering a peek at galaxies farther out. Since the light-bending depends on the mass of the lens, the effect also lets observers weigh the lensing galaxies. They turn out to have far more mass than can be seen. It's part of the universe's mysterious dark matter, the roughly 90 percent of its mass that can't be found in stars, gas, planets, or any other known form of matter.

A cosmic web of dark matter is now thought to have governed where galaxies formed. Dark matter is the universe's hidden architecture, and gravitational lensing is one of the few practical ways to "see" it. An effect Einstein thought insignificant has become a key astronomical tool.

Theorists have also dusted off his discarded cosmological constant to explain a startling new discovery. Einstein's "biggest blunder" is now starting to look like one of his greatest successes. Astronomers had assumed that gravity is gradually slowing the expansion of the universe. But in the late 1990s, two teams, measuring the distances to faraway exploding stars, found just the opposite. Like buoy markers spreading apart on ocean currents, these supernovae revealed that space-time is ballooning outward at an accelerating pace.

For Einstein, the cosmological constant was a way to steady the universe. But if its repulsive effect—now called dark energy—is big enough, it could also drive the acceleration. "The need came back, and the cosmological constant was waiting," says Adam Riess of the Space Telescope Science

Institute, one of the discoverers of the acceleration. "It's totally an Einsteinian concept." So was that other prediction of general relativity recently confirmed, the gravitational waves emitted by the collision of such astronomical heavyweights as black holes and neutron stars (see chapter 20).

The mighty jolt of cosmic birth probably also generated gravity waves, which would still be resonating through the cosmos. These remnant ripples could hold direct evidence of the fleeting moment when physicists believe all of nature's forces were united. If so, Einstein's gravity waves could at last offer clues to something he tried and failed to develop: a "theory of everything." Physicists are still seeking such a theory—a single explanation for both the large-scale force of gravity and the short-range forces inside the atom.

Catching these faint echoes of the Big Bang is a major goal of NASA's next generation of space astronomy missions, a plan the agency has tagged "Beyond Einstein."

Beyond Einstein? Not by a long shot. Einstein might be startled by the universe as we understand it today. But it is unmistakably his.

CHAPTER TWENTY-NINE

The Big Burp

The universe began in warp drive

NEARLY four decades ago Alan Guth, now an MIT physicist, introduced the astounding idea that our universe began not with a bang, but with a sort of cosmic burp—a brief moment of superaccelerated expansion that transformed a subatomic smudge of energy into a cosmos capable of generating galaxies, stars, and planets. Ever since, this idea has been avidly investigated and challenged.

Guth trained in particle physics and had no plans to pursue cosmology until the late 1970s. Then, in 1978, he and a fellow postdoc at Cornell University, Henry Tye, were analyzing theories on the unification of the forces of nature. Guth and Tye wondered whether unification in the very early universe might have given rise to magnetic monopoles: hypothetical particles that have only one magnetic pole, either north or south. Continuing to work together after Guth moved to another postdoc position at Stanford University, the two concluded that, indeed, so many monopoles would have been created in the standard conception of the Big Bang that, as Guth said, "we began to wonder why the universe was here at all. [The monopoles'] tremendous weight would have closed the universe back up eons ago." The monopoles would be so gravitationally attracted to one another, an expansion could never get started. Space-time would have collapsed.

To explain why that didn't happen, the two young researchers surmised that the early universe "supercooled" as it expanded, keeping the forces unified a bit longer as temperatures plunged, just as water can sometimes supercool and remain liquid below its freezing point under certain conditions. According to their calculations, such supercooling would have curbed huge numbers of monopoles from being produced.

MIT physicist Alan Guth.
(Betsy Devine/Wikimedia Commons)

Things really got interesting when Guth decided one night to quickly check how such supercooling might have affected the expansion of the newly born universe. On December 6, 1979, around 11:00 p.m., the young physicist sat down in his makeshift home office and began to work on a series of calculations that within a couple of hours covered four pages. The title at the top of the first page, recorded in small, precise black letters, proclaimed his ambitious intention: he was tackling nothing less than the EVOLUTION OF THE UNIVERSE.

Guth was dealing with the arcane tools of his trade—concepts called "Higgs fields" and "false vacuum states." But, as Guth put down his pen around 1:00 a.m., the bottom line was undeniable. If his equations were valid, the universe did not just expand at the moment of its birth, it tore outward like a fanciful science-fiction spaceship in warp drive. Perhaps inspired by the double-digit rises in the cost of living at the time, Guth came up with an appropriate name for this brief period of hyperacceleration: he called it inflation.

Inflation began around 10^{-35} second into our birth, when the universe was less than a trillionth the size of a proton. Guth saw that the proposed supercooling endowed the universe with a tremendous potential energy, not unlike a rock precariously perched on the edge of a precipice. In this state, gravity, normally a force that draws things together, did a turnabout and became repulsive, causing space-time to balloon outward at a superaccelerated rate for an infinitesimal fraction of a second. But that was enough of a window for our subatomic speck of a cosmos to double in size sixty to a hundred times over. Once inflation ended (when the universe was about the size of a marble or larger), its latent energy was converted into all the particles and radiation that surround us today. It was inflation's demise that actually put the bang into the Big Bang, providing our cosmos with all its necessary building materials. As Guth likes to put it, "The universe is the ultimate free lunch." A lot came out of nearly nothing.

Inflation explained a longtime mystery: the uniformity of the universe from end to end. Caught in an unusual state of expansion, the growing cosmic seed was able to maintain a uniform density as space-time hyperaccelerated outward, so

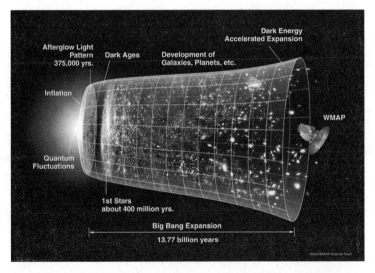

A diagram showing the universe's evolution. The far left depicts the earliest moment, when a period of "inflation" produced a burst of growth. For the next several billion years, space-time continued to expand, more recently speeding up as dark energy came to dominate the expansion.
(NASA/WMAP Science Team)

that our universe ended up looking pretty much the same in all directions. Guth was initially elated by this finding, until he discovered a fatal flaw in his scenario: at the end of his rip-roaring burst, he ended up with a chaotic collection of tiny "bubble universes," none looking like ours. But in the following years, other theorists, such as the Russian physicist Andrei D. Linde, now at Stanford, figured out ways to get one of Guth's many bubbles to balloon into a suitable cosmos.

Yet how do you obtain proof of such a fantastic event, one that occurred at the birth of time itself? If astronomers could peer back with their telescopes to the initial fireball, they wouldn't see anything at all. Much as the Sun's hot outer layers prevent us from gazing to its core, the universe at this time was a blurry soup of plasma, impossible for any optical, radio, or X-ray telescope to probe. The universe didn't become truly transparent until it was about 400,000 years old—when electrons settled down with protons and neutrons to form atoms, and the primordial photons were at last able to travel through the universe unimpeded. Stretched out by the universe's expansion, remnant radiation from the Big Bang now exists as a wash of microwaves bathing the entire universe. Detecting that "cosmic microwave background" tells us how the universe was doing several hundreds of thousands of years after the Big Bang—but no earlier.

Clever theorists, however, found a way around this obstacle. They predicted that quantum fluctuations, tiny jitters in the universe's initial seed, would have blown up to astronomical scales as the universe whizzed outward. And it was those perturbations that helped organize primordial matter into the clusters and galaxies we see today. Valuable support for that idea came when balloons and satellites—sent into space to measure the microwave background with exquisite precision—captured a signal related to temperature with just the pattern of fluctuations predicted by the inflationary models. But competing models for the early universe's behavior, which didn't involve inflation, offered similar predictions.

By the 1990s, theorists offered a more powerful test for inflation—they suggested that primordial gravity waves, generated during inflation, would engrave a unique signature upon the cosmic microwave background. A consequence of Einstein's general theory of relativity, gravity waves are actual ripples in the fabric of space-time, jiggles that alternately stretch and squeeze anything in their path. Searches for this gravity-wave signature were initiated by a number of groups, including teams that have set up special radio telescopes on the icy terrain of the South Pole, notable for its thin, dry air, the best conditions for gathering celestial microwaves (other than in space).

The effect being sought is very subtle. They are looking for a slight swirling pattern on the remnant Big Bang radiation, which indicates it has become "polarized" (the electric fields oscillating back and forth in one preferred orientation). Theory suggests that the gravity waves, as they rippled space-time in the infant universe, had given the light a little kick that caused its orientation to curl, a pattern that only inflationary gravity waves could imprint. Seeing such a pattern would make the case for inflation far stronger. If verified, it's the sort of scientific finding that might prompt its discoverers to think about a Scandinavian vacation in order to pick up their Nobel Prizes.

CHAPTER THIRTY

The Great Escape

"Black holes ain't so black"

GAMMA rays from deep space were discovered by accident in the early 1970s. A group of United States satellites called Vela ("watch" or "vigil" in Spanish) had been put into orbit to make sure nations around the world were complying with the 1963 nuclear test ban treaty. Sifting through the satellites' vast archive of recordings, researchers from the Los Alamos National Laboratory found one event, a burst of gamma rays recorded on July 2, 1967, that didn't look at all like a covert nuclear-bomb test, either in space or on Earth. They soon found similar bursts in the records, and all appeared to come from outside the solar system.

The duration of the bursts ranged from less than a tenth of a second to some thirty seconds—they were popping off like a cosmic flashbulb, flickering for a moment, then fading away. Over the succeeding years various countries launched space detectors that were specifically designed to discern the origin of these powerful cosmic eruptions (gamma rays have the most energy of any electromagnetic radiation), and gradually an answer emerged. Today it's generally accepted that the most common bursts emanate from the gravitational collapse of massive stars—located as far away as the most distant and ancient reaches of the universe—into black holes; others have their origin in collisions between pairs of neutron stars.

Each successive generation of gamma-ray instrumentation offered better and better timing resolution. And that presented physicist David Cline at the University of California, Los Angeles, and several colleagues with a unique opportunity. Plowing through the data from seven gamma-ray detectors, they came to suspect that what they called Very Short Gamma Ray Bursts—those lasting less than a tenth of a second—might represent a class of phenomena with a distinct cause.

How to explain these ultra-brief, super-high-energy bursts? Cline and his colleagues claim they could be evidence for tiny "primordial black holes," perhaps with the mass of a small asteroid packed into the volume of an atomic nucleus, that formed within the extreme densities of the early universe—a phenomenon first predicted by Stephen Hawking in the 1970s. That would be *big* news in the physics community, if true, for such bursts would then offer the means to study what happens when general relativity (the rules that govern the universe at large) merges with quantum mechanics (the tenets of the atomic world). Such a union of the macrocosm with the microcosm has long been sought by physicists.

Hawking's musings were partly sparked during a visit to Moscow in the fall of 1973, where he talked with Soviet physicists Yakov Zel'dovich and Alexander Starbinsky. Those two men had suggested that under special circumstances—that is, when a black hole rotates—it should convert that rotational energy into radiation, thus creating particles. This emission would continue until the spinning black hole wound down and stopped turning.

Devising his own mathematical attack on the problem, Hawking was surprised to discover that all black holes—spinning or not—would be radiating. As Hawking later put it, "Black holes ain't so black."

Hawking announced his discovery in February 1974 at a quantum gravity conference held in England, and his report was soon published in the journal *Nature*. In this endeavor, Hawking looked at the black hole from the perspective of an atom and found that quantum mechanical effects caused black holes to create and emit particles as hot bodies would. As a consequence, the black hole slowly decreases in mass and eventually disappears in a final explosion. Such a finding turned black-hole physics upside-down; a black hole, by definition, holds on to everything it swallows. It's supposed to emit nothing and never go away.

Hawking estimated it would take more than 10^{60} years, far longer than the age of the universe, for a regular black hole, weighing a few stellar masses, to disappear. But what if extremely small holes were created in the turbulence of the Big Bang? They could be popping off right now. Hawking estimated that in its final breath—its last tenth of a second of life—that tiny object would release the energy of a million one-megaton hydrogen bombs.

Needless to say, his fellow physicists were not enthralled by this idea. At that February conference, it was greeted with total disbelief. At the end of Hawking's talk, the chairman of the session, John Taylor from Kings College, London, got up and responded, "Sorry, Stephen, but this is absolute rubbish."

But gradually, over the following two years, it came to be recognized that Hawking had made a startling breakthrough: his argument demonstrated that gravitation and quantum mechanics were somehow deeply connected. Even though these two laws of nature have yet to be fully joined, here was evidence that unity was achievable.

Hawking saw that space-time gets so twisted near a black hole that it enables pairs of particles (a matter particle and its antimatter mate) to pop into existence just outside the black hole. You could think of it as energy being extracted from the black hole's intense gravitational field and then converted into particles.

But because we're talking about the submicroscopic scale, the exact line of the black hole's boundary is quite fuzzy. So,

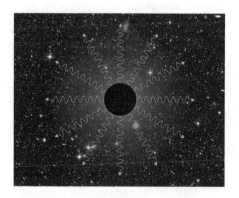

Illustration of a black hole evaporating,
releasing radiation over time.
(APS/Alan Stonebraker)

at times, one of the newly created particles can disappear into the black hole, never to return, while the other remains outside and flies off. As a result, the hole's total mass-energy is reduced a smidgen. This means the black hole is actually evaporating. Ever so slowly, particle by particle, the black hole is losing mass.

While it would take trillions upon trillions of years for a regular black hole to shrink away to nothing, what if the universe did manufacture those multitudes of tiny black holes— mini–black holes—during the first moments of the Big Bang, as Hawking has suggested? Like a ball rolling down a hill, the evaporation of a mini–black hole would accelerate as time progresses. The more mass this tiny primordial object loses, the faster and faster it fizzles away, until it reaches a cataclysmic end.

If the Big Bang did forge such holes, the smallest would have vanished before their dying light could catch our attention; but objects containing the mass of a mountain, yet compressed to the size of a proton, would have continued shedding the last of their mass in brief, spectacular bursts of gamma rays.

That's what Cline and his colleagues believe they might be seeing within the gamma-ray detector records. Others are not so sure. Such signals could also be arriving from a more mundane stellar activity, one not yet identified. As Carl Sagan liked to say, "Extraordinary claims require extraordinary evidence." Cline agrees and is urging other researchers to start studying these events as well, to see if his team's claim holds up to scrutiny. If the distinctive pop of a primordial black hole is at last verified, it will be a significant moment in astronomical history.

CHAPTER THIRTY-ONE

Meet the Multiverse

A theoretical physicist brings her bewildering
science down to Earth

IT began with Sir Isaac Newton. With the publication of
his *Principia* in 1687, Newton became the first scientist
to demonstrate that nature's actions, from the path of a
cannonball to the Moon's orbit about the Earth, could
be described by distinct mathematical laws. Mathematics became the key to unlocking the secrets of the heavens.

Continuing along this path, the Scottish theorist James
Clerk Maxwell in the 1860s devised a concise set of eminently
beautiful equations that united the forces of electricity and
magnetism, showing them to be different sides of the same
coin. Several decades later, Albert Einstein, spurred by his superb physical intuition but also by an astute mathematical
rigor, extended Newton's laws and showed that gravity was a
geometric manifestation. Space-time became a palpable
item—a flexible sheet—and objects that appear to be under a
gravitational force are actually following the geometric curves
that matter impresses upon this rubbery mat of space-time.
Even before tests confirmed this view, Einstein was sure his

theory was right because of what he called "its incomparable beauty."

Mathematical beauty is a potent lure to physicists. In 1963, Murray Gell-Mann looked at the bewildering array of ephemeral particles discovered by physicists and found order by imagining a more fundamental group of building blocks called quarks, which combined by specific rules to generate the many particles. At that time, theoretical physicists were generally working side by side with experimentalists, but, encouraged by their successes, the theorists began to race ahead into unknown territory. The most ambitious, guided solely by the beauty and power of their mathematics, built a construct known as superstrings. This theory suggests that all the forces we experience and the particles we detect result from infinitesimally small strings vibrating within a space-time composed of ten or eleven dimensions.

The story of superstrings was skillfully told in Brian Greene's best-selling *The Elegant Universe*, but there's another aspect to this tale that Greene kept in the background. Not all theoretical physicists are happy with this dependence on mathematical splendor. Some are worried that the notorious celebrity of superstrings has diverted many of the best and brightest in physics from their science's more traditional (and successful) strategy: teaming up with experimentalists. Just as journalists Bob Woodward and Carl Bernstein were advised in the movie *All the President's Men* to "follow the money" to reach their goal—exposing the Watergate scandal—superstring critics would like to see theorists follow the data.

Superstring mavens are the top-downers. They flew up to the ethereal heights and are now looking back down at the real world, hoping to find experimental evidence for strings below them. But, as Harvard professor of physics Lisa Randall asked in *Warped Passages*, have they now found themselves "at the edge of a precipitous, isolated cliff, too remote for them to find their way back to base camp"?

Randall represents the other faction of theorists: those whose feet are firmly planted near an atom smasher and who make predictions that will be either accepted or rejected as particles are slammed together and the resulting debris sifted through. They are the "model builders," who offer a healthy dose of caution to the grander claims of superstring theory. "So far," writes Randall, "all attempts to make string theory realistic have had something of the flavor of cosmetic surgery.

Physicist Lisa Randall.
(Festival della Scienza/Wikimedia Commons)

In order to make its predictions conform to our world, theorists have to find ways to cut away the pieces that shouldn't be there, removing particles and tucking dimensions demurely away. String theory is captivating at first, but ultimately string theorists have to address these fundamental problems." She says the model builders, on the other hand, are the "trailblazers who are trying to find the path that connects the solid ground below to the peak. They yield definite predictions for physical phenomena, giving experimenters a way to verify or contradict a model's claims."

The two camps are not totally at odds. Indeed, Randall acknowledges that the inroads made by string theorists have been inspirational in part for her and her colleagues. "String theory introduces new ideas, both mathematical and physical, that no one would otherwise have considered, such as extra-dimensional notions," she notes.

String theory brought to the forefront the idea that there may be more to the universe than just three spatial dimensions—height, width, length—plus time. There could be six more dimensions that we fail to perceive, possibly because they are so tiny and curled up and hidden from view, or perhaps because some are infinite in extension. These new spatial directions are Randall's "warped passages."

At first, theorists postulated that it was the strings themselves that oscillated within these many dimensions, allegedly creating the various particles found in the cosmos. More recently, that idea has expanded to include membranes, or "branes" for short. A brane is essentially a slice out of that multidimensional world. According to this view, we might be living on a four-dimensional brane (space + time), which itself is immersed inside the full dimensional realm known in its entirety as the "bulk." Such entities as light waves, electrons, and protons are confined to our specific brane, much like water droplets rolling down a shower curtain.

This setup introduces us to a new and mind-blowing take on the universe, or should we say "multiverse." We may be residing amid other branes, other parallel universes, within this complex higher-dimensional domain. "Thinking about branes makes you aware of just how little we know about the space in which we live," says Randall. "The universe might be a magnificent composition linking intermittent branes." If there is life on those other branes, they likely experience different forces and possibly even different forms of matter. Despite the science-fiction quality to this notion, evidence for these higher dimensions might actually be obtainable in the foreseeable future. "Experimental tests of competing hypotheses are near at hand, and within a decade," she predicts, "there should be a dramatic revision in our understanding of fundamental physical laws that will incorporate whatever is discovered."

For the past few decades, many theorists have been focused on unifying the four forces of nature—gravity, electromagnetism, and the strong and weak nuclear forces. Just as Maxwell showed that electricity and magnetism were different features of the same force—electromagnetism—so, too, do theorists suspect that all the forces at some time were united, likely in the first moment of the Big Bang. As the primordial universe cooled and expanded, each force took on its own identity. But there might be more important questions to answer first. Why are the masses of the elementary particles—such entities as the electrons, protons, and neutrons that make

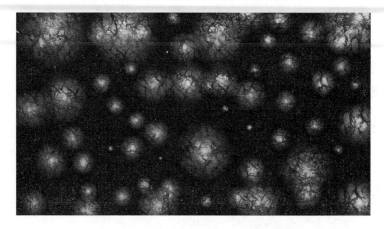

An artistic imagining of the multiverse.
(GiroScience/Shutterstock)

up atoms—so low (theory alone would predict masses much higher), and why is the force of gravity so weak, compared with the other forces? A toy magnet, for instance, can lift a paper clip off the ground, despite the entire Earth gravitationally pulling back on it.

The investigations of higher dimensions by Randall and her fellow model builders are centered upon these conundrums, and they offer several schemes for possible testing. In one model, for example, every particle we know and see around us has a partner in higher dimensional space—a KK particle (named after Theodor Kaluza and Oskar Klein, two physicists who first toyed with the idea of higher dimensions in the early twentieth century). According to Randall, these particles originate in the extra dimensions but make an appearance in our universe with measurable properties. In a way, they cast a three-dimensional "shadow" upon our world, much as an object would cast a two-dimensional shadow on a wall on a sunny day.

Finding a ghostly KK particle would not only be evidence of the higher dimensions, but would also provide an answer to gravity's frailty. Whereas electromagnetism and the nuclear forces are confined to our brane, and so remain fairly strong, gravity could be the lone force that spans all the dimensions and, as a consequence, gets diluted. Or maybe, posits Randall, we live near a brane where gravity is intensely strong, but by the time the gravitational field extends through a fifth dimension, it arrives on our brane of space-time much weakened.

Most exciting for Randall and her colleagues in this endeavor is that testable predictions can be made, renewing the exhilarating time in particle physics of the late 1960s, when quarks were first detected at the Stanford Linear Accelerator as electrons slamming into protons revealed that the protons were built out of three smaller particles (as Gell-Mann had surmised theoretically).

CERN, the European particle-physics center situated on the Swiss-French border, recently installed the most powerful instrument ever built to investigate the properties of elementary particles. Two beams of protons smash together at energy levels so high that the resulting impact might nudge some KK particles into plain sight (or at least allow them to leave their calling cards within the collision debris). What is more, infinitesimally tiny black holes might form as well, quickly evaporating in a hail of energy. There's even a small chance that strings themselves might be amplified and detected. Any of these occurrences would be evidence of higher dimensions.

How seriously should we take all this talk of vibrating strings and parallel universes? Hypotheses in high-energy physics rise and fall on the internet these days, sometimes in a matter of hours. But I can imagine getting comfortable with branes and higher dimensions, as some of us are already accustomed to black holes, relativity, and particle/wave dualities.

CHAPTER THIRTY-TWO

When the Universe Began,
What Time Was It?

To learn how the cosmos blossomed out of a subatomic
point, theorists must first settle a fundamental question: is
time, at the smallest of physical scales, irrelevant?

TIME is an elusive notion. Poets often think of
time as a river, a free-flowing stream that carries
us from the radiant morning of birth to the
golden twilight of old age. It is the span that
separates the delicate bud of spring from the lush flower of
summer.

Physicists think of time in somewhat more practical
terms. For them, time is a means of measuring change—an
endless series of instants that, strung together like beads, turn
an uncertain future into the present and the present into a
definite past. The very concept of time allows researchers to
calculate when a comet will round the Sun or how a signal
traverses a silicon chip. Each step in time provides a peek at
the evolution of nature's myriad phenomena.

In other words, time is a tool. In fact, it was the first scientific tool. Ancient astronomers meticulously tracked the

Sun's march across the Zodiac in order to mark off the seasons and determine when to plant and harvest. In this day and age, solar timepieces have been replaced by atomic clocks that, thanks to the steady pulsing of hydrogen or other atoms, do not gain or lose a second in millions of years. Time can now be sliced into slivers as thin as one ten-trillionth of a second.

But what is being sliced? Unlike mass and distance, time cannot be perceived by our physical senses. We don't see, hear, smell, touch, or taste time. And yet we somehow measure it. Captivated by this conundrum, physicists are beginning to explore the very origins of time. And on first look, they are wondering whether time is a fundamental property of the universe at all. Maybe it is solely a personal experience, set up by our minds to distinguish then from now. As the joke goes, "Time is nature's way of preventing everything from happening all at once."

Such thoughts are more than philosophic. As a cadre of theorists attempt to extend and refine the general theory of relativity, Einstein's momentous law of gravitation, they have a problem with time. A big problem.

"It's a crisis," says mathematician John Baez, of the University of California at Riverside, "and the solution may take physics in a new direction." Not the physics of our everyday world. Stopwatches, pendulums, and hydrogen maser clocks will continue to keep track of nature quite nicely here in our low-energy earthly environs. The crisis arises when physicists attempt to merge the macrocosm—the universe on its grandest scale—with the microcosm of subatomic particles.

Gravity is the weakest of nature's forces, but gravity gains collective strength as masses accumulate and exert their effect over larger and larger distances. The force that causes one object to attract another eventually comes to control the motions of planets, stars, and galaxies. And the best description of how that happens is contained in Einstein's general theory of relativity, introduced in 1915. But the domain in which this theory works is limited; it does not apply to problems at the subatomic scale. For decades, physicists have struggled to discern how gravity acts on the level of elementary particles, a realm governed by the quite different set of rules laid down by quantum mechanics. Arranging this rather curious marriage—an all-embracing theory of "quantum gravity"—is one of physics' last great tasks.

There is a vital reason for physicists' dogged pursuit of this problem. They believe that quantum gravity was the dominant force at the birth of the universe, during the first tiny 10^{-43} second (one ten-millionth of a trillionth of a trillionth of a trillionth of a second). It was an instant when all the matter and energy in the universe was squeezed into a space far smaller than a proton. The microcosm and the macrocosm, in effect, were crushed together in a "singularity," a freakish state where density advances toward infinity and volume approaches zero.

By figuring out the physics of such a bizarre realm, theorists may at last find the key to the origins of the universe, how it came into existence. Simultaneously, they would be learning what lies at the heart of a black hole, the gravitational abyss that is thought to result when the core of an exploding star is crushed inward until its size becomes atomic rather than celestial.

A solution to this mystery, it turns out, lies in understanding the meaning of time: how it acts—and whether it even exists—at the moment of creation or deep within a black hole. Telling time, after all, involves picking out something in the world around you that is changing—the Sun rising and setting, pendulums swinging—and tracking those changes to establish a chronology. With a clock, one can determine the sequence of events; and with a sequence of events, one can properly analyze the behavior of a system—in other words, "do the physics." But how do you register time, the most basic widget in a physicist's toolbox, when the entire mass of a stellar core is squeezed into a subatomic speck? Or when the entire visible universe is in such a state? What kind of clock could physicists possibly use to deal with the crushing and featureless conditions that marked the universe's birth, when quantum gravity was in control?

The problem is really a mathematical one but can be visualized in this crude way: Imagine you could somehow shift a magical gear into reverse and travel back some 14 billion years to that moment of creation. For most of the trip, a wristwatch would work just fine in keeping track of time. But upon reaching the very cauldron of creation, the watch would melt in a nanosecond. You could still keep track of time through the constant vibrations of individual atoms, the basis of atomic clocks. But go back far enough and even atoms cease to exist. Soon there is no longer any means of measuring the progress of events. During that primordial moment when the force of quantum gravity was strongest and the cosmos was tinier than a nuclear particle, there was essentially no room to place a clock, safe from interference, and gauge how the universe was evolving.

This dilemma summarizes the problem of time in physics. Either theorists come up with a "quantum clock," a means of understanding and dealing with the passage of time in that minuscule province where gravity and the quantum world mingle (at least on paper), or they do away with the concept of time altogether.

"The problem of time is one of the deepest issues in physics that must be addressed," says theoretical physicist Christopher Isham of Imperial College in London. And more than timekeeping is at stake here. There will be no Theory of Everything—no peek at "the mind of God," as the Cambridge University cosmologist Stephen Hawking so famously put it in *A Brief History of Time*—until this mystery is resolved. Time plays such an integral role in most laws of physics that physicists are starting to worry: without a sense of time, a definable clock at the moment of creation, will it be possible to explain all of nature's varied forces with one unified law? The question has been lurking in the background, like some crazy relative hidden away in the attic, as physicists seek that Holy Grail.

Time became a key word in the language of physics during the seventeenth century, notably when Isaac Newton wove the passage of time directly into his equations, as in *force = mass ×* *acceleration*. Today, it is difficult for any physicist to examine the universe without thinking of time in much the same way as the illustrious Britisher did more than three hundred years ago. Most of the laws of physics continue to be written in the style of Newton; they are designed to show how things change from one moment to the next. Each event under study, such as the path of a ball thrown into the air or the thermodynamics of a melting ice cube, is broken down into a series of freeze-frames that, run like a movie, show how nature works.

Newton had placed a clock upon the mantel of the universe. This Newtonian timepiece ticked and tocked, chiming like some cosmic Big Ben, in step with all celestial inhabitants, no matter what their speed or position. That meant that a clock situated at the edge of the universe or zipping about the cosmos at high velocities would register the same passage of time—identical minutes and identical seconds—as an earthbound clock. More important, the Newtonian clock was never affected by the events going on around it. Time was aloof and absolute, alike for all as galaxies collided, solar systems formed, and moons orbited planets. Time led an independent existence, separate from nature itself.

This comfortable notion of time held until the beginning of this century, but then it was shattered with a jolt. Albert Einstein uncovered a glitch in Newton's cozy clockwork. With his special theory of relativity, published in 1905, Einstein showed that a clock at rest and a clock in motion do not necessarily agree with one another. Each registers a different flow of time. This effect is well documented: a muon particle (a heavy electron) racing in from space at near the speed of light, for instance, lives many times longer than a muon at rest on Earth. What Einstein did was transform time into a true physical entity, one that was changed by what was going on around it. With special relativity, physicists learned that time is not absolute, as Newton had us think. Time, it turns out, is in the eye of the beholder and in the beholder's surroundings.

Three years after this revelation appeared in print, Einstein's teacher Hermann Minkowski took Newton's clock off the mantelpiece and rolled it out like cookie dough to form the cosmic landscape called *space-time*. Minkowski, wanting to better explain some of special relativity's unusual properties, glued space and time together to form a seamless canvas, a new absolute framework in which time becomes physically connected to space. If you think of space-time coordinates as the interwoven threads of a blanket, tweaking one set of threads will affect all the others: travel near the speed of light and space will shrink as time expands. "Henceforth space by itself, and time by itself, are doomed to fade away into mere shadows," remarked Minkowski. Time alone can no longer be separated from the mix.

In 1915, with his revolutionary general theory of relativity, Einstein shook up the classical, Newtonian view of time even further. He took the novel image of space-time and warped it, and in so doing was able to explain the origin of gravity, long a mystery. According to Newton, rocks fell to Earth and planets orbited the Sun because these objects were somehow held by invisible tendrils of force. Why should this be so? No one knew. But with Einstein's insight, the tendency of one object to attract another object became a simple matter of geometry. It was the natural consequence whenever a mass distorted the space-time canvas. A massive body—the Sun, for example—indents the mat (much the way our bodies can sink into a flexible mattress), and nearby objects must then circle it because they are caught, like cosmic marbles, in the deep space-time basin carved out by the Sun.

General relativity treats time very differently from the way it's handled in other areas of physics. Under Newton, time was special. Every moment was tallied by a universal clock that stood separate and apart from the phenomenon under study. In general relativity, this is no longer true. Einstein declared that time is not absolute—no particular clock is special—and his equations describing how the gravitational force works take this into account. His law of gravity looks the same no matter what timepiece you happen to be using as your gauge. "In general relativity time is completely arbitrary," explains Imperial College's Isham. "The actual physical predictions that come out of general relativity don't depend on your choice of a clock." The predictions will be the same whether you are using a clock traveling near the speed of light or one sitting quietly at home on a shelf.

The choice of clock is still crucial, however, in other areas of physics, particularly quantum mechanics. It plays a central role in Erwin Schrödinger's celebrated wave equation of 1926. The equation shows how a subatomic particle, whether traveling alone or circling an atom, can be thought of as a collection of waves, a wave packet that moves from point to point in space and from moment to moment in time.

According to the vision of quantum mechanics, energy and matter are cut up into discrete bits, called quanta, whose motions are jumpy and blurry. They fluctuate madly. The behavior of these particles cannot be worked out exactly, the way a rocket's trajectory can. Using Schrödinger's wave equation, you can only calculate the probability that a particle—a wave packet—will attain a certain position or velocity. This is a picture so different from the world of classical physics that even Einstein railed against its indeterminacy. He declared that he could never believe that God would play dice with the world.

You might say that quantum mechanics introduced a fuzziness into physics: You can pinpoint the precise position of a particle, but at a trade-off; its velocity cannot then be measured very well. Conversely, if you know how fast a particle is going, you won't be able to know exactly where it is. Werner Heisenberg best summarized this strange and exotic situation with his famous uncertainty principle. But all this action, uncertain as it is, occurs on a fixed stage of space and time, a steadfast arena. A reliable clock is always around—is always needed, really—to keep track of the goings-on and thus enable physicists to describe how the system is changing. At least, that's the way the equations of quantum mechanics are now set up.

And that is the crux of the problem. How are physicists expected to merge one law of physics—namely gravity—that requires no special clock to arrive at its predictions, with the subatomic rules of quantum mechanics, which continue to work within a universal, Newtonian time frame? In a way, each theory is marching to the beat of a different drummer (or the ticking of a different clock).

That's why things begin to go a little crazy when you attempt to blend these two areas of physics. Although the scale on which quantum gravity comes into play is so small that current technology cannot possibly measure these effects directly, physicists can imagine them. Place quantum particles on the springy, pliable mat of space-time, and it will bend and fold like so much rubber. And that flexibility will greatly affect the operation of any clock keeping track of the particles. A timepiece caught in that tiny submicroscopic realm would probably resemble a pendulum clock laboring amid the quivers and shudders of an earthquake. "Here the very arena is being subjected to quantum effects, and one is left with nothing to stand on," explains Isham. "You can end up in a situation where you have no notion of time whatsoever." But quantum calculations depend on an assured sense of time.

For Karel Kuchař (pronounced KOO-cosh), a general relativist and professor emeritus at the University of Utah, the key to measuring quantum time is to devise, using clever math, an appropriate clock—something he has been attempting, off and on, for several decades. Conservative by nature, Kuchař believes it is best to stick with what you know before moving on to more radical solutions. So he has been seeking what might be called the submicroscopic version of a Newtonian clock, a quantum timekeeper that can be used to describe the physics going on in the extraordinary realm ruled by quantum gravity, such as the innards of a black hole or the first instant of creation.

Unlike the clocks used in everyday physics, Kuchař's hypothetical clock would not stand off in a corner, unaffected by what is going on around it. It would be set within the tiny, dense system where quantum gravity rules and would be part and parcel of it. This insider status has its pitfalls: the clock would change as the system changed—so to keep track of time, you would have to figure out how to monitor those variations. In a way, it would be like having to pry open your wristwatch and check its workings every time you wanted to refer to it.

The most common candidates for this special type of clock are simply "matter clocks." "This, of course, is the type of clock we've been used to since time immemorial. All the clocks we have around us are made up of matter," Kuchař points out. Conventional timekeeping, after all, means choosing some material medium, such as a set of particles or a fluid, and marking its changes. But with pen and paper, Kuchař mathematically takes matter clocks into the domain of quantum gravity, where the gravitational field is extremely strong and those probabilistic quantum-mechanical effects begin to arise. He takes time where no clock has gone before.

But as you venture into this domain, says Kuchař, "matter becomes denser and denser." And that's the Achilles heel for any form of matter chosen to be a clock under these extreme conditions; it eventually gets squashed. That may seem obvious from the start, but Kuchař needs to examine precisely how the clock breaks down so he can better understand the process and devise new mathematical strategies for constructing his ideal clock.

More promising as a quantum clock is the geometry of space itself: monitoring space-time's changing curvature as the infant universe expands or a black hole forms. Kuchař surmises that such a property might still be measurable in the extreme conditions of quantum gravity. The expanding cosmos offers the simplest example of this scheme. Imagine the tiny infant universe as an inflating balloon. Initially, its surface bends sharply around. But as the balloon blows up, the curvature of its surface grows shallower and shallower. "The changing geometry," explains Kuchař, "allows you to see that you are at one instant of time rather than another." In other words, it can function as a clock.

Unfortunately, each type of clock that Kuchař has investigated so far leads to a different quantum description, different predictions of the system's behavior. "You can formulate your quantum mechanics with respect to one clock that you place in space-time and get one answer," explains Kuchař.

"But if you choose another type of clock, perhaps one based on an electric field, you get a completely different result. It is difficult to say which of these descriptions, if any, is correct."

More than that, the clock that is chosen must not eventually crumble. Quantum theory suggests there is a limit to how fine you can cut up space. The smallest quantum grain of space imaginable is 10^{-33} centimeter wide, the Planck length, named after Max Planck, inventor of the quantum. (To give you an idea how tiny that is, if an atom were blown up to the size of our Milky Way galaxy, which spans some 100,000 light-years, this quantum grain would still be no bigger than a human cell.) On that infinitesimal scale, the space-time canvas turns choppy and jumbled, like the whitecaps on an angry sea. Space and time become unglued and start to wink in and out of existence in a probabilistic froth. Time and space, as we know them, are no longer easily defined. This is the point at which the physics becomes unknown and theorists start walking on shaky ground. As physicist Paul Davies points out in his book *About Time,* "You must imagine all possible geometries— all possible spacetimes, space warps and timewarps—mixed together in a sort of cocktail, or 'foam.'"

Only a fully developed theory of quantum gravity will show what's really happening at this unimaginably small level of space-time. Kuchař conjectures that some property of general relativity (as yet unknown) will not undergo quantum fluctuations at this point. Something might hold on and not come unglued. If that's true, such a property could serve as the reliable clock that Kuchař has been seeking for so long. And with that hope, Kuchař continues to explore, one by one, the varied possibilities.

Kuchař has been trying to mold general relativity into the style of quantum mechanics, to find a special clock for it. But some other physicists trying to understand quantum gravity believe that the revision should happen the other way around— that quantum gravity should be made over in the likeness of general relativity, where time is pushed into the background. Carlo Rovelli is a champion of this view.

"Forget time," Rovelli declares emphatically. "Time is simply an experimental fact." Rovelli, a physicist at the Center of Theoretical Physics in France, has been working on an approach to quantum gravity that is essentially timeless. To simplify the calculations, he and his collaborators, physicists Abhay Ashtekar and Lee Smolin, set up a theoretical space without a clock. In this way, they were able to rewrite Einstein's general theory of relativity, using a new set of variables so that it could more easily be interpreted and adapted for use on the quantum level.

Their formulation has allowed physicists to explore how gravity behaves on the subatomic scale in a new way. But is that really possible without any reference to time at all? "First with special relativity and then with general relativity, our classical notion of time has only gotten weaker and weaker," answers Rovelli. "We think in terms of time. We need it. But the fact that we need time to carry out our thinking does not mean it is reality."

Rovelli believes if physicists ever find a unified law that links all the forces of nature under one banner, it will be written without any reference to time. "Then, in certain situations," says Rovelli, "as when the gravitational field is not dramatically strong, reality organizes itself so that we perceive a flow that we call time."

Getting rid of time in the most fundamental physical laws, says Rovelli, will probably require a grand conceptual leap, the same kind of adjustment that sixteenth-century scientists had to make when Copernicus placed the Sun, and not the Earth, at the center of the universe. In so doing, the Polish cleric effectively kicked the Earth into motion, even though back then it was difficult to imagine how the Earth could zoom along in orbit about the Sun without its occupants being flung off the surface. "In the 1500s, people thought a moving earth was impossible," notes Rovelli.

But maybe the true rules are timeless, including those applied to the subatomic world. Indeed, a movement has been under way to rewrite the laws of quantum mechanics, a renovation that was spurred partly by the problem of time, among other quantum conundrums. As part of that program, theorists have been rephrasing quantum mechanics' most basic equations to remove any direct reference to time.

The roots of this approach can be traced to a procedure introduced by the physicist Richard Feynman in the 1940s, a method that has been extended and broadened by others, including James Hartle of the University of California at Santa Barbara and physics Nobel laureate Murray Gell-Mann.

Basically, it's a new way to look at Schrödinger's equation. As originally set up, this equation allows physicists to compute the probability of a particle moving directly from point A to point B over specified slices of time. The alternate approach introduced by Feynman instead considers the infinite number of paths the particle could conceivably take to get from A to B, no matter how slim the chance. Time is removed as a factor; only the potential pathways are significant.

Summing up these potentials (some paths are more likely than others, depending on the initial conditions), a specific path emerges in the end. Consider a ball being thrown across a street to your neighbor's house. There's a high probability it will take the shortest and straightest route, but others are possible. The ball could steeply arc, for instance; it could swerve to the right or to the left; there's even a minuscule chance it could go around the Earth in the opposite direction and hit your neighbor's back door. Each path represents a potential outcome for the particle and contributes to the final result.

The process is sometimes compared to interference between waves. When two waves in the ocean combine, they may reinforce one another (leading to a new and bigger wave) or cancel each other out entirely. Likewise, you might think of these many potential paths as interacting with one another—some getting enhanced, others destroyed—to produce the final path. More important, the variable of time no longer enters into the calculations.

Hartle has been adapting this technique to his pursuits in quantum cosmology, an endeavor in which the laws of quantum mechanics are applied to the young universe to discern its evolution. Instead of dealing with individual particles, though, he works with all the configurations that could possibly describe an evolving cosmos, an infinite array of potential universes. When he sums up these varied configurations—some enhancing one another, others canceling each other out—a particular space-time ultimately emerges. In this way, Hartle hopes to obtain clues to the universe's behavior during the era of quantum gravity. Conveniently, he doesn't have to choose a special clock to carry out the physics: time disappears as an essential variable.

Of course, as Isham points out, "having gotten rid of time, we're then obliged to explain how we get back to the ordinary world, where time surrounds us." Quantum gravity theorists have their hunches. Like Rovelli, many are coming to suspect that time is not fundamental at all. This theme resounds again and again in the various approaches aimed at solving the problem of time. Time, they say, may more resemble a physical property such as temperature or pressure. Pressure has no meaning when you talk about one particle or one atom; the concept of pressure arises only when we consider trillions of atoms. The notion of time could very well share this statistical feature. If so, reality would then resemble a pointillist painting. On the smallest of scales—the Planck length—time would have no meaning, just as a pointillist painting, built up from dabs of paint, cannot be fathomed close up. At that range, the painting looks like nothing more than a random array of dots. But as you move back, the dots begin to blend together and a recognizable picture slowly comes into focus. Likewise, space-time, the entity so familiar to us, might take form and reveal itself only when we scrutinize larger and larger scales. Time could be simply a matter of perception, present on the large scale but not on the smallest scale imaginable. Physicists talk of the universe "congealing" or "crystallizing" out of the chaotic quantum jumble that lies at the heart of the Big Bang. Time is not a physical entity but rather a notion that emerges.

Hawking at Cambridge University saw such an effect in his own work on quantum cosmology. To arrive at this conclusion Hawking first had to circumvent the unique and complicated status of time in the space-time continuum. While time can be considered a fourth dimension, it is very different from length, width, and height. In space an object can move freely in any direction—but in time an object must always move forward into the future and away from the past. And this requirement makes the mathematics of quantum cosmology quite complicated. The equations are tough to handle. Hawking decided to get rid of this restriction by treating time as just another dimension of space—a mathematical procedure (trick may be too strong a word) physicists often use to simplify what would otherwise be an intractable problem. The equation has been altered, but its solution can sometimes provide an inkling of the answer hidden in the more complicated equation. In the 1930s, quantum theorists used a similar approach to figure out how radioactive elements can eject subatomic particles. By all the classical laws of physics, the protons and neutrons within an atom don't have enough energy to break free from the steely grip of an atomic nucleus. But physicists keenly grasped that, in the probabilistic world of the atom, there were small but real odds that a particle could acquire enough energy every once in a while to "tunnel" through its nuclear barriers and fly out of the atom.

Hawking's foray into that nebulous realm where general relativity meets quantum mechanics suggested that time, nonexistent at first, could have emerged in an analogous fashion, burrowing into the real world from a domain of timelessness. Thus, there is no reason to inquire what came before the Big Bang. To Hawking, that was as senseless a question as asking what is north of the North Pole.

There's another way to look at Hawking's result: Time simply loses all meaning as you travel back, closer and closer to the Big Bang singularity, akin to the way a compass starts gyrating and loses its ability to indicate a precise direction as you near the north or south magnetic pole. A compass is useful only when it's far from a magnetic pole; likewise, time may be discernible only after you get far enough away from the Big Bang singularity. Perhaps St. Augustine got it right when he wrote, in the fifth century, that "the world was made, not in time, but simultaneously with time."

Unfortunately, St. Augustine did not reveal by what means, and that is the mystery that is so vexing. Hawking's mathematical procedure offered a glimpse, not a final solution. Physicists as yet only recognize the problem, and sense what must happen, but are far from postulating a definitive mechanism. That awaits a full theory of quantum gravity.

Quantum gravity theorists like to compare themselves to archeologists. Each investigator is digging away at a different site, finding a separate artifact of some vast subterranean city. The full extent of the find is not yet realized. What theorists desperately need are data, experimental evidence that could help them decide between the different approaches.

It seems an impossible task, one that would appear to require re-creating the hellish conditions of the Big Bang. But not necessarily. For instance, future generations of "gravity-wave telescopes," instruments that detect ripples in the rubberlike mat of space-time, might someday sense the Big Bang's reverberating thunder, relics from the instant of creation when the force of gravity first emerged. Such waves could provide vital clues to the nature of space and time.

"We wouldn't have believed just [decades] ago that it would be possible to say what happened in the first ten minutes of the Big Bang," points out Kuchař. "But we can now do that by looking at the abundances of the elements. Perhaps if we understand physics on the Planck scale well enough, we'll be able to search for certain consequences—remnants—that are observable today." If found, such evidence would bring us the closest ever to our origins and possibly allow us to perceive at last how space and time came to well up out of nothingness some 14 billion years ago.

Notes

CHAPTER 1. EARTH IS BUT A SPECK

This chapter was first published in the *Washington Post*, Bartusiak (July 2009d).

CHAPTER 2. BEDAZZLED BY A COMET

nine times in the past three centuries: Rao (2012).

"that the Soul of Caesar": Secundus (1847–48), p. 65.

"the most difficult of the whole book": Newton (1999), p. 270.

first comet to be discovered with a telescope: Levy (1998), p. 12.

"comets are a kind of planet": Newton (1999), p. 895.

"solid, compact, fixed, and durable": Ibid., p. 918.

"extremely thin vapor": Ibid., p. 919.

suggested that threads of magnetic force: Discussed in Kepler (1995).

planets carried around like leaves: Discussed in Descartes (1998).

"I have not as yet been able to deduce": Newton (1999), p. 943.

"The space between the Sun and the fixed stars": Halley (1705), p. 20.

"I dare venture to foretell": Ibid., p. 22.

The comet appeared on schedule: Levy (2003), p. 26.

CHAPTER 3. TO BE . . . OR NOT TO BE A PLANET

They looked for five years: Bartusiak (1996), pp. 46, 49.

proposed . . . by, among others, the Dutch-American astronomer Gerard P. Kuiper: Jewitt (n.d.).

"The confirmation of the Kuiper belt changes our perception": Bartusiak (1996), p. 50.

Michael E. Brown announced in 2005: Brown, Trujillo, and Rabinowitz (2005).

dubbed Eris: Brown and Schaller (2007).

Pluto was demoted: Overbye (2006).

astronomers sought an underlying pattern: Littmann (1988), p. 14, and Hoskin (1999), pp. 158–59.

The pattern from Bode's *On the New Eighth Major Planet* (table): "Bode and Piazzi" (1929).

"Can one believe that the Creator of the Universe": Hoskin (1999), p. 159.

jokingly referred to itself as the "celestial police": Ibid., p. 160.

"The light was a little faint, and of the color of Jupiter": Piazzi (1801). Also in Bartusiak (2004), p. 151.

"Since its movement is so slow": Abetti (1974), p. 592.

Carl Friedrich Gauss was able to calculate its orbit: Littman (1988), p. 19.

"Piazzi had, indeed, here discovered a very extraordinary object": "Bode and Piazzi" (1929), p. 182. Also in Bartusiak (2004), p. 151.

Ceres was smaller than our Moon: Herschel (1802).

he suggested the name *asteroid:* Ibid., p. 228.

Today it is known they are a field of debris: Hoskin (1999), p. 162.

CHAPTER 4. THE WATERY ALLURE OF MARS

Mars-shattering, discovery: Jerolmack (2013) and Williams et al. (2013).

strengthen the case that liquid water once flowed freely: Jerolmack (2013), p. 1056.

He was one of the Boston Brahmins: Strauss (2001), p. 3.

"After lying dormant for many years": Lowell (1935), p. 5.

"is not without a considerable atmosphere": Herschel (1784), p. 273.

"Considerable variations observed in the network of waterways": Pannekoek (1989), p. 378.

adding 116 waterways to Schiaparelli's original depiction: Lane (2006), p. 199.

"a trial to sane astronomers": Sheehan (1996), p. 132.

"the surface of Mars was (and still is) notoriously difficult to make out":
  Lane (2006), p. 201.
a few dark markings were seen: Ibid., p. 205.
news story of the year: "Mars" (1907).

CHAPTER 5. RINGS, RINGS, RINGS

due to the transit of a giant ringed planet: See Kenworthy and Mamajek
  (2015).
"very strange wonder": Van Helden (1974), p. 105.
"The star of Saturn is not a single star": Ibid.
"in three minor knots divided": Hall (2014), p. 1318.
"a strange metamorphosis": Deiss and Nebel (1998), p. 216.
"Saturn deceives or really mocks": Van Helden (1974), p. 108.
Saturn looked as if it had handles: Ibid., p. 110.
"the problem of Saturn's appearances had become a celebrated puzzle":
  Ibid., p. 115.
"arms extended on both sides": "Classics of Science" (1929), p. 191.
these arms had vanished altogether: Van Helden (1974), p. 120.
"surrounded by a thin flat ring": Pollack (1975), p. 3.
"pure fiction": Brashear (1999).
solid structure would be highly unstable: Pollack (1975), p. 4.
In a prize-winning 1856 essay: Maxwell (1859).
And that's exactly what Keeler measured: Osterbrock (2002), pp. 158–64;
  Keeler (1895).

CHAPTER 6. THE BAFFLING WHITE DWARF STAR

had enough data to announce that Sirius and Procyon were not traveling smoothly: Bessel (1844).
completed one orbit every fifty years: Ibid., p. 139.
"The subject . . . seems to me so important": Ibid., p. 136.
and thus lost in the glare: Holberg and Wesemael (2007), p. 167.
"there might have been a [prearranged] connection": Welther (1987).
"It remains to be seen": Bond (1862), p. 287.
Clark in 1862 garnered the prestigious Lalande Prize: Holberg and
  Wesemael (2007), pp. 170–71.
Russell doubted that such a classification could be correct: Holberg
  (2007), p. 114.

Walter Adams at the Mount Wilson Observatory in California confirmed the spectrum: Adams (1915).

"I was flabbergasted": Philip and DeVorkin (1977).

"The message of the companion of Sirius": Eddington (1927), p. 50.

British theorist Ralph Fowler finally figured out: Fowler (1926).

"star of large mass . . . cannot pass into the white-dwarf stage": Chandrasekhar (1934), p. 377.

"there should be a law of nature": "Meeting of the Royal Astronomical Society" (1935), p. 38.

"continued gravitational contraction:" Oppenheimer and Snyder (1939).

"Only its gravitational field persists": Ibid., p. 456.

CHAPTER 7. THE STAR NO BIGGER THAN A CITY

"No event in radio astronomy seemed more astonishing": Hey (1973), p. 139.

his calculations indicated that the dwarf would undergo further stellar collapse: Chandrasekhar (1931).

in a spectacular stellar explosion they had christened a "supernova": Baade and Zwicky (1934b), p. 254.

Baade first referred to them as *Hauptnovae*: Osterbrock (2001), p. 32.

"forming one gigantic nucleus": Landau (1932), p. 288.

would transform completely into naked spheres of neutrons: Baade and Zwicky (1934a), p. 263. It should be noted that astronomers later learned that there are essentially two types of supernovae: one, called Type II, involves a massive star's core collapsing to either a neutron star or black hole; the other, Type I, is when a white dwarf steals gas from a companion. If enough matter is stolen, the dwarf star ignites in a runaway reaction that blows up the star.

only a handful of physicists . . . proceeded to investigate a neutron star's possible structure: See Gamow (1937), Oppenheimer and Serber (1938), and Oppenheimer and Volkoff (1939).

"there is about as little hope of seeing such a faint object": Wheeler (1964), p. 195.

"I like to say that I got my thesis with sledgehammering": Bartusiak (1986), p. 42.

"there was a little bit of what I call 'scruff'": Kellermann and Sheets (1983), pp. 164–65.

"I was [then] two-and-a-half years through a three-year studentship": Ibid., p. 168.

"lots of little green men on opposite sides of the universe": Interview of Jocelyn Bell Burnell by David DeVorkin on May 21, 2000, Niels Bohr Library & Archives, American Institute of Physics, College Park, Maryland, www.aip.org/history-programs/niels-bohr-library/oral-histories/31792.

the news was finally released in February 1968: Hewish et al. (1968).

"One of [the photographers] even had me running down the bank": Bell Burnell (1977), p. 688.

dubbed the novel objects *pulsars:* "Anthony Michaelis" (2008).

"likened to radio bursts from a solar flare": Hewish et al. (1968), p. 712.

Thomas Gold developed the model that best explained a pulsar's behavior: Gold (1968).

at least a few hundred million neutron stars now reside in the Milky Way: Camenzind (2007), p. 269.

Hewish had been skeptical about Bell's "scruff": Interview of Jocelyn Bell Burnell by David DeVorkin on May 21, 2000, Niels Bohr Library & Archives, American Institute of Physics, College Park, Maryland, www.aip.org/history-programs/niels-bohr-library/oral-histories/31792.

Nobel now stood for "No Bell": Ibid.

"I believe it would demean Nobel Prizes": Bell Burnell (1977), p. 688.

CHAPTER 8. YE OLD BLACK HOLE

Michell was a geologist, astronomer, mathematician, and theorist: Details of his life can be found in McCormmach (1968).

"the most inventive of the eighteenth-century natural philosophers": Ibid., p. 127.

"father of modern seismology": Hardin (1966), p. 30.

"short Man, of a black Complexion": Ibid., p. 27.

"the odds against the contrary opinion": Michell (1767), p. 249.

"arguably the most innovative and perceptive contribution": Montgomery, Orchiston, and Whittingham (2009), p. 91.

he began monitoring and cataloging the stars positioned close together: Herschel (1782).

Michell decided to extend his ideas on double stars: Michell (1784).

Michell was devoted to the Society: Jungnickel and McCormmach (1999), p. 565, note 7.

some historians have speculated: Ibid., p. 564, and Montgomery, Orchiston, and Whittingham (2009), p. 91.

"diminution of the velocity": Michell (1784), p. 35.

"all light . . . would be made to return": Ibid., p. 42.

"A luminous star of the same density as the Earth": Laplace (1809), p. 367.

he expunged his invisible-star speculation: Gillispie (1997), p. 175.

CHAPTER 9. AS THOUGH NO OTHER NAME EVER EXISTED

In June 1756, on the banks of the Hooghly river: Details of the Black Hole of Calcutta come from Cavendish (2006).

"Well, after I used that phrase four or five times": Bartusiak (2000), p. 62.

their existence remained a well-kept secret: Interview with Joseph Taylor at Texas Symposium on Relativistic Astrophysics, December 2013.

Wheeler's name is missing from the official conference proceedings: See Brancazio and Cameron (1969).

The term then made it into print: Wheeler (1968).

"Gravitational collapse would result": Rosenfeld (1964), p. 11.

is sure he didn't invent the term: Phone interview with Rosenfeld, 2012.

"space may be peppered with 'black holes'": Ewing (1964), p. 39.

originated the term *quasar*: Chiu (1964), p. 21.

"To the astonished audience, he jokingly added": A letter dated May 25, 2009, describing Chiu's knowledge on the origin of the term "black hole" was sent by Chiu to *Physics Today*. It was not published, but Chiu kindly provided a copy to me.

His sons told McHugh: An email from John Dicke to Loyala University physicist Martin McHugh, with the kind permission of both to use it.

"He simply started to use the name": Thorne (1994), p. 256.

needed to be held at a distance within quotation marks: See Kafka (1969) and Sullivan (1968).

"He accused me of being naughty": Wheeler and Ford (1998), p. 297.

"Thus *black hole* seems the ideal name": Ibid.

"The advent of the term black hole": Wheeler (1990), p. 3.

CHAPTER 10. LIKE THIS WORLD OF OURS

In 2017 an international team of astronomers thrillingly revealed: Gillon et al. (2017).

"infinite worlds both like and unlike": Oates (1940), p. 5.

seemed logical that they'd ultimately construct: Dick (1998), p. 8.

"never be perceived by us": Herschel (1791), p. 74.

faster than any other star: Barnard (1916).

van de Kamp got worldwide attention: van de Kamp (1963).

failed to confirm the Barnard-star finding: Gatewood and Eichhorn (1973).

Bruce Campbell and Gordon Walker pioneered a way: Campbell and Walker (1979).

two momentous events: Aumann et al. (1984) and B. Smith and Terrile (1984).

"probably consists of 'second generation' planets": Wolszczan and Frail (1992), p. 147.

That long-anticipated event: Mayor and Queloz (1995).

They first revealed their discovery: "51 Pegasi" (1995).

"spectacular detection": Marcy and Butler (1996), p. L147.

Other discoveries followed swiftly: Ibid.

"reminiscent of solar system planets": Butler and Marcy (1996), p. L153.

a trio of planets: Butler et al. (1999).

"We've gone from the early days": Chu (2017).

"that stars are orbited by planets as a rule": Cassan et al. (2012), p. 169.

CHAPTER 11. OUR SPIRALING HOME

one of the best baby pictures of the cosmos: Overbye (2013).

At first they tried just counting stars: Gingerich (1985), p. 59.

designated an "enemy alien": Osterbrock (2001), p. 98.

Baade came to recognize that highly luminous blue and blue-white supergiant stars: Ibid., p. 102.

He first teamed up with Jason Nassau: Gingerich (1985), p. 64.

the two students set up a special camera: Ibid., p. 68.

provided the breakthrough: See Morgan, Sharpless, and Osterbrock (1952).

used cotton balls to depict: "Spiral Arms of the Galaxy" (1952).

"Astronomers are usually of a quiet and introspective disposition": Struve (1953), p. 277.

clapping of hands: Gingerich (1985), p. 69.

"finally found its solution": Ibid.

Within two years, the spiraling segments were confirmed: Van de Hulst, Muller, and Oort (1954).

our galaxy has two dominant arms: Churchwell et al. (2009), p. 228.

CHAPTER 12. THE WOMAN WHO CHASED GALAXIES

he could see out to distances of a few hundreds of millions of light-years: Hubble (1936), p. 517.

"There, we measure shadows": Hubble (1982), p. 202.

"are enormous systems": Hubble (1936), p. 543.

a choice that hadn't been available to her in New Zealand: Biographical details on Beatrice Tinsley are largely provided by Eisberg (2001).

"one of the boldest graduate thesis projects": Kennicutt (1999), p. 1165.

constructed her model based on the best theoretical and observational evidence: Tinsley (1968).

"Brian Tinsley's clever wife": Eisberg (2001), p. 268.

What they saw matched Tinsley's prediction: Butcher and Oemler (1978).

"changed the course of cosmological studies": Faber (1981), p. 110.

No longer able to use her right hand: Larson and Stryker (1982), p. 165.

was published the following November: Tinsley (1981).

CHAPTER 13. STUFF OF THE HEAVENS

In 2006 the probe returned to Earth's vicinity: Leary (2006).

"The significance of this discovery": Brownlee (2009).

"we are made of starstuff": Sagan (1980), p. 190.

"we would never know how to study": Hearnshaw (2014), p. 1.

"But people would say we must have gone mad": "Some Scientific Centres" (1902), p. 587.

he had turned his spectroscope to the heavens: Kirchhoff (1862), pp. 20–21.

Within a few years, other astronomers: Hearnshaw (2014), pp. 36–37, 41–44.

"It is remarkable that the elements": Huggins and Miller (1864), p. 434.

In 1955, physicist Charles H. Townes ... was invited to address an international symposium: Bartusiak (1993), pp. 169–70.

would be "hopeless": Ibid., p. 173.

they found hydroxyl radicals screaming out: Weinreb et al. (1963).

recorded the radio cries of both ammonia and water: Cheung et al. (1968) and Cheung et al. (1969).

handed out cases of liquor: Bartusiak (1993b), p. 175.

forms for every 30 million molecules of hydrogen: Ibid., p. 174.

Hydrogen peroxide, the hair-bleaching agent, was uncovered: Bergman et al. (2011).

CHAPTER 14. RECIPE FOR THE STARS

This chapter was first published in *The Sciences*, Bartusiak (1993a): Marcia Bartusiak, "The Stuff of Stars." *The Sciences*, September/October 1993a, pp. 34–39. It draws on a number of sources, including Haramundanis (1984), DeVorkin (1989), DeVorkin and Kanat (1983a,b), Kidwell (1990), Payne (1925), and an interview of Jesse Greenstein by Spencer R. Weart on April 7, 1977, Niels Bohr Library & Archives, American Institute of Physics, College Park, Maryland, www.aip.org/history-programs/niels-bohr-library/oral-histories/4643-1.

CHAPTER 15. FIND A WAY AROUND IT

This chapter was first published in *Smithsonian* magazine, Bartusiak (2005b). Copyright 2005 Smithsonian Institution. Reprinted with permission from Smithsonian Enterprises. All rights reserved. Reproduction in any medium is strictly prohibited without permission from Smithsonian Institution.

One of the most important discoveries: Burbidge et al. (1957).

B²FH delivered the proof: Bartusiak (2004), pp. 366–68.

beheld them with vivid clarity: Burbidge (1994), p. 3.

astronomy could be a career: Ibid., p. 4.

using a telescope so antiquated: Ibid., p. 7.

she fought for—and won—access: Ibid., p. 18.

"Thanks to her influence": http://womensmuseumca.org/hall-of-fame/ margaret-burbidge. Accessed September 22, 2017.

"If you meet with a blockage": Burbidge (1994), p. 9.

a feat listed in the *Guinness Book of Records:* http://cwp.library.ucla.edu/ Phase2/Burbidge,_E._Margaret@932123456.html. Accessed September 22, 2017.

first female director of the famed Royal Greenwich Observatory: Burbidge (1994), p. 30.

has not always embraced the majority opinion: Email interview with Burbidge, July 18, 2005.

"continually surprised by the almost religious fervor": Burbidge (1994), p. 26.

more attracted to the notion . . . that matter was created in successive epochs: Ibid., p. 35.

"To ride with the telescope": Ibid., p. 25.

CHAPTER 16. DARK MATTERS

They serve the Cryogenic Dark Matter Search: Cho (2013).

dense object that he called a "neutron star": Baade and Zwicky (1934a), p. 263.

"It is difficult to understand": Zwicky (1937b), p. 234.

referred to this invisible substance as *dunkle Materie:* Zwicky (1933), p. 125.

largely owing to Vera Rubin: Bartusiak (1990), pp. 91–92.

To their surprise, they revealed that the stars and gas: Rubin, Ford, and Thonnard (1980).

Modeling this effect, theorists figured: Ostriker and Peebles (1973).

by 1978 Rubin and her team had measured more than two hundred: Bartusiak (1990), p. 94.

aimed the Hubble Space Telescope at massive galaxy clusters: Coe et al. (2010).

there is five times as much dark matter: According to Bennett et al. (2013), p. 46, normal matter composes 4.628 percent of the universe's total mass-energy, while the dark matter density is 24.02 percent. Other measurements have slightly different values but are in the same range.

CHAPTER 17. COSMIC FUNHOUSE

astronomers from Great Britain, Russia, and Spain announced in 2009: Belokurov et al. (2009).

racing toward one another at around 300,000 miles . . . per hour: Irwin et al. (2015).

thrust Einstein into the public eye: "Lights All Askew in the Heavens" (1919).

Eddington considered the possibility: Eddington (1920), pp. 133–35.

"as to make it impossible to detect it": Ibid., p. 134.

its light would spread out to form a ring: Chwolson (1924).

Einstein was already aware: Renn, Sauer, and Stachel (1997).

"Some time ago, [Rudi] W. Mandl"; "a most curious effect"; "no hope": Einstein (1936).

"little value, but it makes the poor guy": Renn, Sauer, and Stachel (1997), p. 186.

"Extragalactic nebulae [galaxies] offer"; "see [other] nebulae": Zwicky (1937a).

confirmed that the cozy pair were quasars: Walsh, Carswell, and Weymann (1979).

The lens turned out to be a giant elliptical: Young et al. (1980).

"The vistas we uncover with this new gravitational telescope": Gates (2009), p. 5.

CHAPTER 18. RIVERS OF GALAXIES

dubbed this gargantuan structure "Laniakea": Tully et al. (2014), p. 73.

named it (rather uninspiredly) the "Local Group": Hubble (1982), p. 125.

"remarkable collection": Herschel (1785), pp. 255–56.

"metagalactic systems" or "metagalactic clouds": See Holmberg (1937) and Shapley (1933).

To him, galactic groupings stopped at clusters: Hubble (1982), p. 187.

de Vaucouleurs had been an expert observer of Mars: Interview of Gerard de Vaucouleurs by Alan Lightman on November 7, 1988,

Niels Bohr Library & Archives, American Institute of Physics, College Park, Maryland, www.aip.org/history-programs/niels-bohr-library/oral-histories/33930.

He called it the "Local Supergalaxy": de Vaucouleurs (1953).

"It was considered as sheer speculation": Bartusiak (1986), p. 170.

"All of a sudden": Biviano (2000), p. 9.

Abell . . . pointed out other potential superclusters: Abell (1961).

"striking confirmation": Ibid., p. 610.

as if they are on the surfaces of huge, nested bubbles: De Lapparent, Geller, and Huchra (1986).

CHAPTER 19. THE BIG DIPPER IS CRYING

"All we see is a blob in the sky": Siegel (2014).

a high-tech wonder: Details on the array can be found at http://www.telescopearray.org.

Did this ionization originate from the Earth's crust: Carlson (2013), p. 10.

built a sensitive electroscope: Ibid., pp. 10–11.

started taking measurements aboard balloons: Ibid., p. 11.

"radiation coming from above": Ibid., p. 12, and Walter (2012), p. 25.

"the whole of the penetrating radiation": Otis and Millikan (1924), p. 778.

convinced of their extraterrestrial nature: Carlson (2013), p. 13.

"found wild rays more powerful": "Millikan Rays" (1925).

he called them "cosmic rays": De Maria and Russo (1989), p. 214.

"signals broadcasted throughout the heavens": Millikan (1928), pp. 281, 282–83.

the national press regularly covered this scientific tussle: Carlson (2013), p. 14.

The particle model finally won in 1932: Compton (1933).

could also be atomic nuclei or electrons: Jones (2013), p. 17.

physicists came to discover new and bizarre elementary particles: Walter (2012), pp. 38–39.

rays came from spectacular stellar blasts: Baade and Zwicky (1934a).

CHAPTER 20. EINSTEIN'S SYMPHONY

a gravitational wave passed through the Earth on September 14, 2015: Abbott et al. (2016).

He predicted that a pair of masses: Einstein (1916).

He claimed to have observed such ringing: Weber (1969).

Weiss wrote a landmark report: Weiss (1972).

construct a pair of large detectors with arms two-and-a-half-miles . . . long: Bartusiak (2017), pp. 141–43.

physics community quickly protested: Ibid., p. 147.

The wave first arrived at Livingston: Details of the first detection in Bartusiak (2017), pp. 177–204.

assumed it was a "blind injection": Interview with Marco Drago, January 15, 2016.

Could it have been a hacker?: Interview with Gabriela Gonzalez, January 8, 2016.

One of the holes weighed thirty-six solar masses: Abbott et al. (2016).

"the warped side of the universe": Interview with Kip Thorne, January 8, 2016.

It was almost guaranteed: All the succeeding descriptions of the various gravitational-wave events come from Bartusiak (2017), pp. 226–39.

"That is the prize": LIGO news conference, Massachusetts Institute of Technology, Boston, Massachusetts, February 11, 2016.

CHAPTER 21. UNDERGROUND ASTRONOMY

Deep beneath the South Pole, thousands of detectors: See https://icecube.wisc.edu.

The ultrahigh energy of this special set of particles: Aartsen et al. (2015).

Chadwick was soon sent to an internment camp: Details on Chadwick's and Ellis's wartime experiences can be found in Sutton (1992), p. 14, and Hutchison, Gray, and Massey (1981), pp. 201–2.

"Dear radioactive ladies and gentlemen": Sutton (1992), p. 7.

"desperate": Ibid., p. 21.

Chadwick discovered the first known electrically chargeless particle: Chadwick (1932).

Fermi dubbed Pauli's hypothetical particle the neutrino: Fermi (1934), p. 161.

"the little one who was not there": Ne'eman and Kirsh (1986), p. 71.

to stop one in its tracks: Ibid., p. 73.

set up a detector outside a South Carolina nuclear power plant: Ibid., p. 73.

"All good things come to the man": Enz (2002), p. 488.

Raymond Davis set up the first neutrino observatory: See https://www. bnl.gov/bnlweb/raydavis/research.htm.

CHAPTER 22. EAVESDROPPING ON THE UNIVERSE

the VLA, for Very Large Array: Further details on its operation can be found at http://www.vla.nrao.edu.

"by far the most sensitive such radio telescope": Finley (2012).

assigned to investigate long-radio-wave static: Friis (1965), p. 841.

"Jansky's merry-go-round": See http://www.nrao.edu/whatisra/hist_ jansky.shtml.

Jansky at last established in 1932 that the disruptive 20-megahertz static: Jansky (1933).

dubbed the signal his "star noise": Friis (1965), p. 842.

not the "result of some form of intelligence": "New Radio Waves Traced" (1933).

NBC's public affairs–oriented Blue Network broadcast the signal: Kellerman and Sheets (1983), p. 47.

"sounded like steam escaping": "Radio Waves Heard from Remote Space" (1933).

"some sort of thermal agitation": Jansky (1935), p. 1162.

violent streams of electrons spiraling about: Kellerman and Sheets (1983), p. 55.

infant field at last took off: Reich and Wielebinski (2002).

Long burdened with a chronic kidney ailment: Kellerman and Sheets (1983), p. 40.

In his last experiments, he was trying out a newfangled gadget: Friis (1965), p. 842.

CHAPTER 23. THE ONCE AND FUTURE QUASAR

announced the discovery of the most distant quasar: Bañados et al. (2017).

recognized the first quasar: Bartusiak (1986), p. 151.

Ryle reported that he counted more far-off cosmic radio sources: Ryle (1958).

narrow down the location of a particularly strong source, labeled 3C 48: Matthews and Sandage (1963), p. 30–31.

"I took a spectrum the next night": Thorne (1994), p. 335.

couldn't even find evidence that hydrogen: Hawking and Israel (1989), p. 243.

On the fifth day of that month: Bartusiak (1986), p. 151.

3C 273 was rushing away from us: Schmidt (1963).

made the cover of *Time* magazine: *Time*, March 11, 1966.

"The insult was not that they radiate": Bartusiak (1986), p. 152.

checked old photographic plates: Hawking and Israel (1989), p. 246.

CHAPTER 24. FINDING A COSMIC YARDSTICK

In the early 1890s, the Harvard College Observatory established a southern station: Jones and Boyd (1971), pp. 289–92.

Pickering shrewdly recognized the value of smart young women: Pickering (1898), p. 4.

These woman "computers" . . . photographic magnitude: Jones and Boyd (1971), pp. 388–90.

began work as a volunteer soon after graduating: Many of the details of Leavitt's life are drawn from George Johnson's excellent biography of Henrietta Leavitt. See Johnson (2005).

found a record-setting total of 1,777 new variable stars: Leavitt (1908).

"It is worthy of notice": Ibid., p. 107.

She had found her law: Leavitt and Pickering (1912).

advised by her doctor to avoid the chilly night air: Johnson (2005), p. 31.

observatory's prime function was to collect and classify data: Jones and Boyd (1971), p. 369.

dedicated herself for several years to a separate project: Johnson (2005), pp. 56–57.

Hertzsprung picked up where she left off: Hertzsprung (1914).

divulged her interest to . . . Harlow Shapley: Harvard University Archives, letter from Shapley to Leavitt, May 22, 1920.

nominate her for a Nobel Prize in Physics: Johnson (2005), p. 118.

CHAPTER 25. THE COSMOLOGIST LEFT BEHIND

This chapter was first published in *Sky & Telescope*, Bartusiak (2009b).

ordered a custom-built spectrograph: Hall (1970b), p. 162.

many of America's greatest astronomers . . . red and blue ends of the spectrum: Smith (1994), pp. 45–48.

eventually becoming a virtuoso . . . made of interstellar dust: Hoyt (1996), pp. 129–45.

"Dear Mr. Slipher, I would like to have you take": Lowell Observatory Archives, Lowell to Slipher, February 8, 1909.

"I do not see much hope of our getting the spectrum": Lowell Observatory Archives, Slipher to Lowell, February 26, 1909.

"This plate of mine": Lowell Observatory Archives, Slipher to Lowell, December 3, 1910.

"It is not really very good": Lowell Observatory Archives, Slipher to Lowell, September 26, 1912.

high-voltage induction coil: Hall (1970a), p. 85.

"encouraging results or (I should say) indications": Lowell Observatory Archives, Slipher to Lowell, December 19, 1912.

On a seeing-quality scale from 1 to 10: Lowell Observatory Archives, Douglass to Lowell, January 14, 1895.

December 29–31 observation details: Lowell Observatory Archives, Spectrogram Record Book II, September 24, 1912, to July 28, 1913, pp. 69–70.

Slipher chose to publish a brief account: Slipher (1913).

"It looks as if you had made a great discovery": Lowell Observatory Archives, Lowell to Slipher, February 8, 1913.

"Spectrograms of spiral nebulae are becoming more laborious": Lowell Observatory Archives, Slipher Papers, Hoyt-V.M. Box, Report F4, titled "Spectrographic Observations of Nebulae and Star Clusters."

"telescopic object of great beauty": Lowell Observatory Archives, Slipher Working Papers, Box 4, Folder 4-4.

"no less than three times that of the great Andromeda Nebula": Ibid.

"When I got the velocity of the Andr. N. I went slow": Lowell Observatory Archives, Slipher to Miller, May 16, 1913.

"I leaned against it": Hall (1970a), p. 85.

"It seems to me, that with this discovery": Lowell Observatory Archives, Hertzsprung to Slipher, March 14, 1914.

"It is a question in my mind": Lowell Observatory Archives, Slipher to Hertzsprung, May 8, 1914.

confident of what he was seeing: Interview of Henry Giclas by Robert
    Smith on August 12, 1987, Niels Bohr Library & Archives, Ameri-
    can Institute of Physics, College Park, Maryland, www.aip.org/
    history-programs/niels-bohr-library/oral-histories/5022.
"about 25 times the average stellar velocity": Slipher (1915), p. 23.
his fellow astronomers rose to their feet: Smith (1982), p. 19.
spirals might be "scattering": Slipher (1917b), p. 407.
to establish that the nebula was indeed a separate island universe:
    Hubble (1925).
identified a mathematical trend in the flight of the galaxies: Hubble
    (1929).
When Dutch astronomer Willem de Sitter . . . casually referred to
    several astronomers: de Sitter (1930), p. 169.
"I consider the velocity-distance relation": Hubble Papers, Henry Hun-
    tington Library, San Marino, California, Hubble to de Sitter, Au-
    gust 21, 1930.
"I regard such first steps as by far the most important of all": Lowell
    Observatory Archives, Hubble to Slipher, March 6, 1953.
"emerged from a combination of radial velocities measured by Slipher":
    Hubble (1953), p. 658.

CHAPTER 26. THE PRIMEVAL ATOM

This chapter was first published in *Technology Review*, Bartusiak (2009c).

"exceptionally brilliant": Harvard University Archives, Eddington to
    Shapley, May 3, 1924.
Lemaître traveled to the United States: Details on his studies there
    can be found in Kragh (1987), pp. 118–19, and Kragh (1990),
    p. 542.
find him just by following the sound of his full, loud laugh: McCrea
    (1990), p. 204.
Two models were already in circulation: Einstein (1917) and de Sitter
    (1917).
"combine the advantages of both": Lemaître (1931c), p. 483.
"a cosmical effect of the expansion of the universe": Ibid., p. 489.
A similar solution, conceived independently in 1922 by the Russian
    mathematician Aleksandr Friedmann: Friedmann (1922).

"Your calculations are correct, but your physical insight is abominable": Smith (1990), p. 57.

"It remains to find the cause": Lemaître (1931c), p. 489.

"unique quantum": Lemaître (1931a).

recoiled from any suggestion that his primeval atom had been inspired by the biblical story of Genesis: Kragh (1990), p. 542, and Kragh (2007), pp. 152–53.

Godart brought confirmation: Deprit (1984), p. 391.

CHAPTER 27. PROVING THE BIG BANG

first appeared in a 1948 scientific paper almost as an afterthought: Alpher and Herman (1948).

disintegrated within their theoretical computations: See Bethe (1939).

George Gamow simply looked around for another locale: Alpher and Herman (2001), p. 20.

Their first report on this mathematical recipe, a one-page synopsis: Alpher, Bethe, and Gamow (1948).

joined by fellow lab employee Robert Herman: Alpher and Herman (2001), p. 72.

"the first thoroughly modern analysis": Weinberg (1977), p. 124.

"the temperature in the universe at the present time": Alpher and Herman (1948), p. 775.

didn't link it to cosmology at all: Alpher and Herman (2001), p. 118.

"did not know that they ought to try": Weinberg (1977), p. 127.

holding a press conference to generate attention: D'Agnese (1999), p. 65.

didn't resurface until the mid-1960s: Dicke et al. (1965); Doroshkevich and Novikov (1964).

"But we should not indulge in sermonizing": Alpher and Herman (2001), p. 122.

CHAPTER 28. IT'S NOW EINSTEIN'S UNIVERSE

This chapter was originally published in *National Geographic*, Bartusiak (2005a).

On January 29, 1931, the world's premier physicist: "Einstein Guest at Mount Wilson" (1931) and Christianson (1995), pp. 205–6.

"Well, my husband does that on the back of an old envelope": Clark (1971), p. 434.

"describes how our universe was born": Bartusiak (2005a), p. 116.

"It does not seem that something like that can exist!": Fölsing (1997), p. 46, and Schilpp (1959), p. 53.

Einstein's special theory of relativity: Einstein (1905).

"The idea is amusing and enticing": Fölsing (1997), p. 196.

British astronomers actually measured this warping: Dyson, Eddington, and Davidson (1920).

Einstein was the first to try: Einstein (1917). He was prompted to do this after a discussion of general relativity with Willem de Sitter in the fall of 1916. Kragh (2007), p. 131.

"as required by the fact of the small velocities of the stars": Translated in Lorentz, Einstein, Minkowski, and Weyl (1923), p. 188.

"The red shift of distant nebulae has smashed my old construction": "Red Shift of Nebulae a Puzzle, Says Einstein" (1931).

biggest blunder: This is not a direct quote from Einstein. The Russian-American physicist George Gamow relayed this story in his autobiography, saying Einstein used the now-famous phrase while they were having a chat one day. Gamow (1970), p. 44.

"The evolution of the world can be compared to a display of fireworks": Lemaître (1950), p. 78. He introduced the idea in a series of papers published in *Nature*. See Lemaître (1931a,b).

"The notion of a beginning ... is repugnant": Eddington (1931), pp. 449–50.

"Answering those questions": Bartusiak (2005a), p. 120.

A birthday cake for the universe would require around 14 billion candles: Freedman (2001). The nearly 14-billion-year age of the universe was also pegged by measurements of the cosmological microwave background. See Bennett et al. (2003, 2013).

space-time is ballooning outward at an accelerating pace: Riess et al. (1998) and Perlmutter et al. (1999).

"The need came back, and the cosmological constant was waiting": Bartusiak (2005a), p. 121.

CHAPTER 29. THE BIG BURP

not with a bang, but with a sort of cosmic burp: Guth (1981).

"we began to wonder why the universe was here at all": Bartusiak (1986), pp. 241–42.

he called it inflation: Ibid., pp. 243–44.

"The universe is the ultimate free lunch": Guth (1997), p. 15.

figured out ways to get one of Guth's many bubbles to balloon: Waldrop (1981), p. 122.

CHAPTER 30. THE GREAT ESCAPE

a burst of gamma rays recorded on July 2, 1967: Klebesadel, Strong, and Olson (1973).

they came to suspect that . . . with a distinct cause: Cline et al. (2011).

convert that rotational energy into radiation: Hawking and Israel (1989), p. 264.

"Black holes ain't so black": Hawking (1988), p. 99.

his report was soon published in the journal *Nature:* Hawking (1974).

release the energy of a million one-megaton hydrogen bombs: Ibid., p. 30.

"Sorry, Stephen, but this is absolute rubbish": Boslough (1985), p. 70.

shedding the last of their mass: Hawking (1974), p. 30.

"Extraordinary claims require extraordinary evidence": Harris (1977), p. 1.

CHAPTER 31. MEET THE MULTIVERSE

This chapter was first published in *Harvard Magazine*, Bartusiak (2005c), as a review of physicist Lisa Randall's book *Warped Passages* (Randall 2005): Marcia Bartusiak, "Meeting the Multiverse." *Harvard Magazine*, November/December 2005, pp. 19–22.

"at the edge of a precipitous, isolated cliff": Randall (2005), p. 73.

"all attempts to make string theory realistic": Ibid., p. 70.

"trailblazers who are trying . . . or contradict a model's claim": Ibid., pp. 71, 72.

"String theory introduces new ideas": Ibid., p. 295.

"Thinking about branes makes you aware": Ibid., p. 60.

"Experimental tests of competing hypotheses": Ibid., p. 242.

CHAPTER 32. WHEN THE UNIVERSE BEGAN,
WHAT TIME WAS IT?

This chapter was first published in *Technology Review*, Bartusiak (1995). It draws on a number of sources, including Coveney and Highfield (1992), Davies (1995), Hawking (1988), Isham (1993, 1994), Kuchař (1992, 1993), Penrose (1989), Rovelli (1993).

"It's a crisis": Bartusiak (1995), p. 56.

"The problem of time is one of the deepest issues": Isham (1993), p. 160.

"the mind of God": Hawking (1988), p. 175.

"Henceforth space by itself, and time by itself": Minkowski (1923), p. 75. This was originally presented as an address to the Eightieth Assembly of German Natural Scientists and Physicians, Cologne, Germany, September 21, 1908.

"In general relativity time is completely arbitrary": Bartusiak (1995), p. 58.

"Here the very arena is being subjected to quantum effects": Ibid., p. 59.

"This, of course, is the type of clock we've been used to": Ibid.

"matter becomes denser and denser": Ibid., p. 60.

"The changing geometry allows you to see": Ibid.

"You can formulate your quantum mechanics": Ibid.

"You must imagine all possible geometries": Davies (1995), p. 181.

"Forget time": Bartusiak (1995), p. 60.

"First with special relativity and then with general relativity": Bartusiak (1995), pp. 60–61.

"Then, in certain situations": Ibid., p. 61.

"In the 1500s, people thought": Ibid.

"having gotten rid of time": Ibid., p. 62.

"the world was made, not in time": Augustine (1998), p. 456.

"We wouldn't have believed": Bartusiak (1995), p. 63.

Bibliography

Aartsen, M. G., et al. "Evidence for Astrophysical Muon Neutrinos from the Northern Sky with IceCube." *Physical Review Letters* 115, 081102, August 21, 2015.

Abbasi, R. U., et al. "Indications of Intermediate-Scale Anisotropy of Cosmic Rays with Energy Greater than 57 EeV in the Northern Sky Measured with the Surface Detector of the Telescope Array Experiment." *Astrophysical Journal Letters* 790 (2014): L21–L25.

Abbott, B. P., et al. "Observation of Gravitational Waves from a Binary Black Hole Merger." *Physical Review Letters* 116, 061102, February 12, 2016.

Abell, George O. "Evidence Regarding Second-Order Clustering of Galaxies and Interactions Between Clusters of Galaxies." *Astronomical Journal* 66 (1961): 607–13.

Abetti, Giorgio. "Giuseppe Piazzi." *Dictionary of Scientific Biography*, vol. 10. New York: Scribner's, 1974.

Adams, Walter S. "The Spectrum of the Companion of Sirius." *Publications of the Astronomical Society of the Pacific* 27 (1915): 236–37.

Alpher, Ralph A., Hans Bethe, and George Gamow. "The Origin of Chemical Elements." *Physical Review* 73 (1948): 803–4.

Alpher, Ralph A., and Robert Herman. "Evolution of the Universe." *Nature* 162 (1948): 774–75.

Alpher, Ralph A., and Robert Herman. *Genesis of the Big Bang.* Oxford: Oxford University Press, 2001.

"Anthony Michaelis." *Daily Telegraph*, March 28, 2008.

Augustine, Saint. *The City of God Against the Pagans*, ed. and trans. R. W. Dyson. Cambridge: Cambridge University Press, 1998.

Aumann, H. H., et al. "Discovery of a Shell Around Alpha Lyrae." *Astrophysical Journal* 278 (1984): L23–L27.

Baade, W., and F. Zwicky. "Cosmic Rays from Super-Novae." *Proceedings of the National Academy of Sciences* 20 (1934a): 259–63.

Baade, W., and F. Zwicky. "On Super-Novae." *Proceedings of the National Academy of Sciences* 20 (1934b): 254–59.

Bañados, Eduardo, et al. "An 800-Million-Solar-Mass Black Hole in a Significantly Neutral Universe at a Redshift of 7.5." *Nature*, published online December 6, 2017. doi: 10.1038/nature25180.

Barnard, E. E. "A Small Star with Large Proper Motion." *Astronomical Journal* 29 (1916): 181–83.

Barrow, John D. *The World Within the World.* Oxford: Oxford University Press, 1988.

Bartusiak, Marcia. *Thursday's Universe.* New York: Times Books, 1986.

Bartusiak, Marcia. "The Woman Who Spins the Stars." *Discover*, October 1990, pp. 88–94.

Bartusiak, Marcia. "The Stuff of Stars." *The Sciences*, September/October 1993a, pp. 34–39.

Bartusiak, Marcia. *Through a Universe Darkly.* New York: HarperCollins, 1993b.

Bartusiak, Marcia. "When the Universe Began, What Time Was It?" *Technology Review*, November/December 1995, pp. 54–63.

Bartusiak, Marcia. "The Remarkable Odyssey of Jane Luu." *Astronomy*, February 1996, pp. 46–49.

Bartusiak, Marcia. *Archives of the Universe.* New York: Pantheon, 2004.

Bartusiak, Marcia. "Beyond the Big Bang: Einstein's Evolving Universe." *National Geographic*, May 2005a, pp. 115–21.

Bartusiak, Marcia. "Margaret Burbidge." *Smithsonian*, November 2005b, pp. 34–35.

Bartusiak, Marcia. "Meeting the Multiverse." *Harvard Magazine*, November/December 2005c, pp. 19–22.

Bartusiak, Marcia. *The Day We Found the Universe*. New York: Pantheon, 2009a.

Bartusiak, Marcia. "The Cosmologist Left Behind." *Sky & Telescope*, September 2009b, pp. 30–35.

Bartusiak, Marcia. "Before the Big Bang." *Technology Review*, September/October 2009c, pp. M14–M15.

Bartusiak, Marcia. "You're Not the Center of the Universe, You Know." *Washington Post*, July 19, 2009d, p. B4.

Bartusiak, Marcia. *Black Hole*. New Haven: Yale University Press, 2015.

Bartusiak, Marcia. *Einstein's Unfinished Symphony*. New Haven: Yale University Press, 2017.

Bell Burnell, S. Jocelyn. "Petit Four." *Annals of the New York Academy of Sciences* 302 (1977): 685–89.

Belokurov, V., et al. "Two New Large-Separation Gravitational Lenses from SDSS." *Monthly Notices of the Royal Astronomical Society* 392 (2009): 104–12.

Bennett, C. L., et al. "First-Year Wilkinson Microwave Anisotropy Probe (WMAP) Observations: Preliminary Maps and Basic Results." *Astrophysical Journal Supplement* 148 (2003): 1–27.

Bennett, C. L., et al. "Nine-Year *Wilkinson Microwave Anisotropy Probe* (*WMAP*) Observations: Final Maps and Results." *Astrophysical Journal Supplement* 208 (2013): 1–53.

Bergman, P., et al. "Detection of Interstellar Hydrogen Peroxide." *Astronomy & Astrophysics* 531 (2011): L8–L11.

Bessel, F. W. "On the Variation of Proper Motions of Procyon and Sirius." *Monthly Notices of the Royal Astronomical Society* 6 (1844): 136–41.

Bethe, Hans A. "Energy Production in Stars." *Physical Review* 55 (1939): 434–56.

Biviano, Andrea. "From Messier to Abell: 200 Years of Science with Galaxy Clusters." arXiv.org, October 20, 2000, https://arxiv.org/abs/astro-ph/0010409.

"Bode and Piazzi." In *A Source Book in Astronomy*, ed. Harlow Shapley and Helen E. Howarth. New York: McGraw-Hill, 1929. Translated from J. E. Bode, *Von dem Neuen, zwischen Mars und Jupiter entdeckten achten Hauptplaneten des Sonnensystems* [On the new eighth major planet discovered between Mars and Jupiter]. Berlin, 1802.

Bond, George P. "On the Companion of Sirius." *American Journal of Science* 33 (1862): 286–87.

Boslough, John. *Stephen Hawking's Universe.* New York: W. Morrow, 1985.

Brancazio, Peter J., and A. G. W. Cameron eds. *Supernovae and Their Remnants: Proceedings of the Conference on Supernovae, Held at the Goddard Institute for Space Studies, NASA 1967.* New York: Gordon and Breach Science Publishers, 1969.

Brashear, Ronald. "Christiaan Huygens and His Systema Saturnium." sil.si.edu, May 1999, http://www.sil.si.edu/DigitalCollections/HST/Huygens/huygens-introduction.htm. Accessed May 25, 2017.

Brown, M. E., and E. L. Schaller. "The Mass of Dwarf Planet Eris." *Science* 316 (2007): 1585.

Brown, M. E., C. A. Trujillo, and D. L. Rabinowitz. "Discovery of a Planetary-Sized Object in the Scattered Kuiper Belt." *Astrophysical Journal Letters* 635 (2005): L97–L100.

Brownlee, Don. "Stardust: A Mission with Many Scientific Surprises." Jet Propulsion Laboratory, October 29, 2009, https://stardust.jpl.nasa.gov/news/news116.html. Accessed May 31, 2017.

Burbidge, E. Margaret. "Watcher of the Skies." *Annual Review of Astronomy and Astrophysics* 32 (1994): 1–36.

Burbidge, E. Margaret, G. R. Burbidge, William A. Fowler, and F. Hoyle. "Synthesis of the Elements in Stars." *Reviews of Modern Physics* 29 (1957): 547–650.

Butcher, Harvey, and Augustus Oemler Jr. "The Evolution of Galaxies in Clusters. I. ISIT Photometry of Cl 0024 + 1654 and 3C 295." *Astrophysical Journal* 219 (1978): 18–30.

Butler, R. Paul, et al. "Evidence for Multiple Companions to ν Andromedae." *Astrophysical Journal* 526 (1999): 916–27.

Butler, R. Paul, and Geoffrey W. Marcy. "A Planet Orbiting 47 Ursae Majoris." *Astrophysical Journal Letters* 464 (1996): L153–L156.

Camenzind, Max. *Compact Objects in Astrophysics: White Dwarfs, Neutron Stars and Black Holes.* Berlin: Springer, 2007.

Campbell, B., and G. A. H. Walker. "Precision Radial Velocities with an Absorption Cell." *Publications of the Astronomical Society of the Pacific* 91 (1979): 540–45.

Carlson, Per. "Discovery of Cosmic Rays." *AIP Conference Proceedings* 1516 (2013): 9–16.

Cassan, A., et al. "One or More Bound Planets per Milky Way Star from Microlensing Observations." *Nature* 481 (2012): 167–69.

Cavendish, Richard. "The Black Hole of Calcutta." *History Today*, June 2006, pp. 60–61.

Chadwick, J. "Possible Existence of a Neutron." *Nature* 129 (1932): 312.

Chandrasekhar, S. "The Highly Collapsed Configurations of a Stellar Mass." *Monthly Notices of the Royal Astronomical Society* 91 (1931): 456–66.

Chandrasekhar, S. "Stellar Configurations with Degenerate Cores." *The Observatory* 57 (1934): 373–77.

Chandrasekhar, S. "The Highly Collapsed Configurations of a Stellar Mass (Second Paper)." *Monthly Notices of the Royal Astronomical Society* 95 (1935): 207–25.

Cheung, A. C., et al. "Detection of NH_3 Molecules in the Interstellar Medium by Their Microwave Emission." *Physical Review Letters* 21 (1968): 1701–5.

Cheung, A. C., et. al. "Detection of Water in Interstellar Regions by Its Microwave Radiation." *Nature* 221 (1969): 626–28.

Chiu, Hong-Yee. "Gravitational Collapse." *Physics Today* 17 (1964): 21–34.

Cho, Adrian. "Dark-Matter Mystery Nears Its Moment of Truth." *Science* (2013): 418.

Cho, Adrian, and Yudhijit Bhattacharjee. "First Wrinkles in Spacetime Confirm Cosmic Inflation." *Science* 343 (2014): 1296–97.

Christianson, Gale E. *Edwin Hubble: Mariner of the Nebulae.* Chicago: University of Chicago Press, 1995.

Chu, Jennifer. "Scientists Make Huge Dataset of Nearby Stars Available to Public." MIT News, February 13, 2017, http://news.mit.edu/2017/dataset-nearby-stars-available-public-exoplanets-0213.

Churchwell, Ed, et al. "The *Spitzer*/GLIMPSE Surveys: A New View of the Milky Way." *Publications of the Astronomical Society of the Pacific* 121 (2009): 213–30.

Chwolson, O. "Über eine mögliche Form fiktiver Doppelsterne [Regarding a possible form of fictitious double stars]." *Astronomische Nachrichten* 221 (1924): 329.

Clark, R. W. *Einstein*. New York: World Publishing Company, 1971.

"Classics of Science: Huygens Explains Saturn's Rings." *Science News Letter* 16 (1929): 191–92.

Cline, David B., et al. "Do Very Short Gamma Ray Bursts Originate from Primordial Black Holes? Review." *International Journal of Astronomy and Astrophysics* 1 (2011): 164–72.

Coe, Dan, et al. "A High-Resolution Mass Map of Galaxy Cluster Substructure: LensPerfect Analysis of A1689." *Astrophysical Journal* 723 (2010): 1678–702.

Compton, Arthur H. "A Geographic Study of Cosmic Rays." *Physical Review* 43 (1933): 387–403.

Couper, Heather, and Nigel Henbest. *New Worlds: In Search of Planets*. Reading, Mass.: Addison-Wesley, 1986.

Coveney, Peter, and Roger Highfield. *The Arrow of Time*. New York: Ballentine Books, 1992.

D'Agnese, Joseph. "The Last Big Bang Man Left Standing." *Discover*, July 1999, pp. 60–67.

Davies, Paul. *About Time*. New York: Simon & Schuster, 1995.

Deiss, Bruno M., and Volker Nebel. "On a Pretended Observation of Saturn by Galileo." *Journal for the History of Astronomy* 29 (1998): 215–20.

De Lapparent, Valérie, Margaret J. Geller, and John P. Huchra. "A Slice of the Universe." *Astrophysical Journal Letters* 302 (1986): L1–L5.

De Maria, M., and A. Russo. "Cosmic Ray Romancing." *Historical Studies in the Physical and Biological Sciences* 19 (1989): 211–66.

Deprit, A. "Monsignor Georges Lemaître." In *The Big Bang and Georges Lemaître*, ed. A. Berger. Dordrecht, Netherlands: D. Reidel, 1984.

Descartes, René. *The World and Other Writings*. Trans. and ed. Stephen Gaukroger. Cambridge: Cambridge University Press, 1998.

de Sitter, W. "On Einstein's Theory of Gravitation, and Its Astronomical Consequences. Third Paper." *Monthly Notices of the Royal Astronomical Society* 78 (1917): 3–28.

de Sitter, W. "On the Magnitudes, Diameters and Distances of the Extragalactic Nebulae, and Their Apparent Radial Velocities." *Bulletin of the Astronomical Institutes of the Netherlands* 5 (1930): 157–71.

de Vaucouleurs, Gérard. "Evidence for a Local Supergalaxy." *Astronomical Journal* 58 (1953): 30–32.

DeVorkin, David H. "Henry Norris Russell." *Scientific American*, May 1989, pp. 126–33.

DeVorkin, David H., and Ralph Kenat. "Quantum Physics and the Stars (I): The Establishment of a Stellar Temperature Scale." *Journal for the History of Astronomy* 14 (1983a): 102–32.

DeVorkin, David H., and Ralph Kenat. "Quantum Physics and the Stars (II): Henry Norris Russell and the Abundances of the Elements in the Atmospheres of the Sun and Stars." *Journal for the History of Astronomy* 14 (1983b): 180–221.

Dick, Steven J. *Life on Other Worlds*. Cambridge: Cambridge University Press, 1998.

Dicke, R. H., et al. "Cosmic Blackbody Radiation." *Astrophysical Journal* 142 (1965): 414–19.

Doroshkevich, A. G., and I. D. Novikov. "Mean Density of Radiation in the Metagalaxy and Certain Problems in Relativistic Cosmology." *Soviet Physics Doklady* 9 (1964): 111.

Dyson, F. W., A. S. Eddington, and C. Davidson. "A Determination of the Deflection of Light by the Sun's Gravitational Field, from Observations Made at the Total Eclipse of May 29, 1919." *Philosophical Transactions of the Royal Society of London, Series A* 220 (1920): 291–333.

Eddington, Arthur. *Space, Time, and Gravitation*. Cambridge: Cambridge University Press, 1920.

Eddington, Arthur. *Stars and Atoms*. Oxford: Clarendon Press, 1927.

Eddington, Arthur. "The End of the World: From the Standpoint of Mathematical Physics." *Nature* 127 (1931): 447–53.

Einstein, Albert. "Zur Elektrodynamik beweger Körper [On the electrodynamics of moving bodies]." *Annalen der Physik* 17 (1905): 891–921.

Einstein, Albert. "Näherungsweise Integration der Feldgleichungen der Gravitation [Approximate integration of the field equations of gravitation]." *Sitzungsberichte der Königlich Preussichen Akademie der Wissenschaften*, June 22, 1916, pp. 688–96.

Einstein, Albert. "Kosmologische Betrachtungen zur allgemeinen Relativitätstheorie [Cosmological considerations arising from the general theory of relativity]." *Sitzungsberichte der Königlich Preußischen Akademie er Wissenschaften zu Berlin* 6 (1917): 142–52.

Einstein, Albert. "Lens-Like Action of a Star by the Deviation of Light in the Gravitational Field." *Science* 84 (1936): 506–7.

Einstein, Albert. "Stationary Systems with Spherical Symmetry Consisting of Many Gravitating Masses." *Annals of Mathematics* 40 (1939): 922–36.

"Einstein Guest at Mount Wilson." *Los Angeles Times*, January 30, 1931, p. A1.

Eisberg, Joann. "Making a Science of Observational Cosmology: The Cautious Optimism of Beatrice Tinsley." *Journal for the History of Astronomy* 32 (2001): 263–78.

Enz, Charles P. *No Time to Be Brief: A Scientific Biography of Wolfgang Pauli*. Oxford: Oxford University Press, 2002.

Ewing, Ann. " 'Black Holes' in Space." *Science News Letter* 85 (1964): 39.

Faber, Sandra. "Beatrice Tinsley." *Physics Today* September 1981, pp. 110–12.

Fermi, E. "Versuch einer Theorie der ß-Strahlen [Experiment of a theory of beta rays]." *Zeitschrift für Physik* 88 (1934): 161–77.

"51 Pegasi." *Central Bureau for Astronomical Telegrams*, Circular no. 6251, October 25, 1995.

Finley, Dave. "Iconic Telescope Renamed to Honor Founder of Radio Astronomy." www.nrao.edu, January 10, 2012, http://www.nrao.edu/pr/2012/jansky.

Fölsing, Albrecht. *Albert Einstein: A Biography*. New York: Viking, 1997.

Fowler, Ralph H. "On Dense Matter." *Monthly Notices of the Royal Astronomical Society* 87 (1926): 114–22.

Freedman, W. L., et al. "Final Results from the Hubble Space Telescope Key Project to Measure the Hubble Constant." *Astrophysical Journal* 553 (2001): 47–72.

Friedmann, A. "Über die Krümmung des Raumes [About the curvature of space]." *Zeitschrift für Physik* 10 (1922): 377–86.

Friis, Harold. "Karl Jansky: His Career at Bell Telephone Laboratories." *Science* (1965): 841–42.

Gamow, G. *Atomic Nuclei and Nuclear Transformations*. Oxford: Oxford University Press, 1937.

Gamow, G. *My World Line*. New York: Viking Press, 1970.

Gates, Evalyn. *Einstein's Telescope*. New York: Norton, 2009.

Gatewood, George, and Heinrich Eichhorn. "An Unsuccessful Search for a Planetary Companion of Barnard's Star (BD + 4 °3561)." *Astronomical Journal* 78 (1973): 769–76.

Gillispie, Charles Coulston. *Pierre-Simon Laplace, 1749–1827: A Life in Exact Science*. Princeton, N.J.: Princeton University Press, 1997.

Gillon, Michaël, et al. "Seven Temperate Terrestrial Planets Around the Nearby Ultracool Dwarf Star TRAPPIST-1." *Nature* 542 (2017): 456–60.

Gingerich, Owen. "The Discovery of the Spiral Arms of the Milky Way." In *The Milky Way Galaxy: Proceedings of the 106th Symposium, Groningen, Netherlands, May 30–June 3, 1983*, pp. 59–70. Dordrecht: Reidel Publishing, 1985.

Gold, T. "Rotating Neutron Stars as the Origin of the Pulsating Radio Sources." *Nature* 218 (1968): 731–32.

Guth, Alan H. "Inflationary Universe: A Possible Solution to the Horizon and Flatness Problems." *Physical Review D* 23 (1981): 347–56.

Guth, Alan. *The Inflationary Universe*. New York: Basic Books, 1997.

Hall, Crystal. "Galileo, Poetry, and Patronage." *Renaissance Quarterly*, Winter 2014, pp. 1296–1331.

Hall, J. S. "V. M. Slipher's Trailblazing Career." *Sky & Telescope*, February 1970a, pp. 84–86.

Hall, J. S. "Vesto Melvin Slipher." *Year Book of the American Philosophical Society*. Philadelphia: American Philosophical Society, 1970b, pp. 161–66.

Halley, Edmond. *A Synopsis of the Astronomy of Comets*. London: Printed for John Senex, 1705.

Haramundanis, Katherine, ed. *Cecilia Payne-Gaposchkin: An Autobiography and Other Recollections*. Cambridge: Cambridge University Press, 1984.

Hardin, Clyde. "The Scientific Work of the Reverend John Michell." *Annals of Science* 22 (1966): 27–47.

Harris, Art. "Second View: Sagan on 'Encounters.'" *Washington Post*, December 16, 1977, pp. 1, 3.

Hawking, Stephen. "Black Hole Explosions?" *Nature* 248 (1974): 30–31.

Hawking, Stephen. *A Brief History of Time*. New York: Bantam Books, 1988.

Hawking, Stephen, and Werner Israel, eds. *Three Hundred Years of Gravitation*. Cambridge: Cambridge University Press, 1989.

Hearnshaw, J. B. *The Analysis of Starlight: Two Centuries of Astronomical Spectroscopy*. Cambridge: Cambridge University Press, 2014.

Herschel, William. "Catalogue of Double Stars." *Philosophical Transactions of the Royal Society of London* 72 (1782): 112–62.

Herschel, William. "On the Remarkable Appearances at the Polar Regions of the Planet Mars." *Philosophical Transactions of the Royal Society of London* 74 (1784): 233–73.

Herschel, William. "On the Construction of the Heavens." *Philosophical Transactions of the Royal Society of London* 75 (1785): 213–66.

Herschel, William. "On Nebulous Stars, Properly So Called." *Philosophical Transactions of the Royal Society of London* 81 (1791): 71–88.

Herschel, William. "Observations on the Two Lately Discovered Celestial Bodies." *Philosophical Transactions of the Royal Society of London* 92 (1802): 213–32.

Hertzsprung, E. "Über die räumliche Verteilung der Veränderlichen vom δ Cephei-Typus [On the spatial distribution of variables of the δ Cephei type]." *Astronomische Nachrichten* 196 (1914): 201–8.

Hewish, Antony, S. Jocelyn Bell, John D. H. Pilkington, Paul Frederick Scott, and Robin Ashley Collins. "Observation of a Rapidly Pulsating Radio Source." *Nature* 217 (1968): 709–13.

Hey, J. S. *The Evolution of Radio Astronomy*. New York: Science History Publications, 1973.

Holberg, Jay B. *Sirius: Brightest Diamond in the Night Sky*. Berlin: Springer, 2007.

Holberg, J. B., and F. Wesemael. "The Discovery of the Companion of Sirius and Its Aftermath." *Journal of the History of Astronomy* 38 (2007): 161–74.

Holmberg, Erik. "A Study of Double and Multiple Galaxies." *Annals of the Observatory of Lund* 6 (1937): 3–173.

Hoskin, Michael. *The Cambridge Concise History of Astronomy*. Cambridge: Cambridge University Press, 1999.

Hoyt, W. G. "Vesto Melvin Slipher." In *Biographical Memoirs*, vol. 52. Washington, D.C.: National Academy Press, 1980.

Hoyt, W. G. *Lowell and Mars*. Tucson: University of Arizona Press, 1996.

Hubble, Edwin P. "Cepheids in Spiral Nebulae." *Publications of the American Astronomical Society* 5 (1925): 261–64.

Hubble, Edwin P. "A Relation Between Distance and Radial Velocity Among Extra-Galactic Nebulae." *Proceedings of the National Academy of Sciences* 15 (1929): 168–73.

Hubble, Edwin. "Effects of Redshifts on the Distribution of Nebulae." *Astrophysical Journal* 84 (1936): 517–54.

Hubble Edwin. "The Law of Redshifts." *Monthly Notices of the Royal Astronomical Society* 113 (1953): 658–66.

Hubble, Edwin. *The Realm of the Nebulae*. New Haven: Yale University Press, 1982.

Huggins, William, and W. A. Miller. "On the Spectra of Some of the Fixed Stars." *Philosophical Transactions* 154 (1864): 413–35.

Hutchison, Kenneth, J. A. Gray, and Harrie Massey. "Charles Drummond Ellis." *Biographical Memoirs of Fellows of the Royal Society* 27 (1981): 199–233.

Irwin, Jimmy A., et al. "The Cheshire Cat Gravitational Lens: The Formation of a Massive Fossil Group." *Astrophysical Journal* 806 (2015): 268–81.

Isham, C. J. "Canonical Quantum Gravity and the Problem of Time." In *Integrable Systems, Quantum Groups, and Quantum Field Theories*, vol. 409. Ed. A. Ibort and M. A. Rodriguez. Dordrecht: Springer, 1993.

Isham, C. J. "Prima Facie Questions in Quantum Gravity." In *Canonical Gravity: From Classical to Quantum. Lecture Notes in Physics*, vol. 434. Ed. J. Ehlers and H. Friedrich. Berlin: Springer, 1994.

Jansky, Karl. "Electrical Disturbances Apparently of Extraterrestrial Origin." *Proceedings of the Institute of Radio Engineers* 21 (1933): 1387–98.

Jansky, Karl. "A Note on the Source of Interstellar Interference." *Proceedings of the Institute of Radio Engineers* 23 (1935): 1158–63.

Jerolmack, Douglas J. "Pebbles on Mars." *Science* 340 (2013): 1055–56.

Jewitt, David. "Why 'Kuiper' Belt?" http://www2.ess.ucla.edu/~jewitt/kb/gerard.html. Accessed May 23, 2017.

Johnson, George. *Miss Leavitt's Stars*. New York: W. W. Norton, 2005.

Jones, Bessie Zaban, and Lyle Gifford Boyd. *The Harvard College Observatory: The First Four Directorships, 1839–1919*. Cambridge, Mass.: Harvard University Press, 1971.

Jones, Lawrence W. "High Energy Physics in Cosmic Rays." *AIP Conference Proceedings* 1516 (2013): 17–22.

Jungnickel, Christa, and Russell McCormmach. *Cavendish: The Experimental Life*. Cranbury, N.J.: Bucknell, 1999.

Kafka, P. "Discussion of Possible Sources of Gravitational Radiation." *Mitteilungen der Astronomischen Gesellschaft* 27 (1969): 134–38.

Keeler, J. E. "A Spectroscopic Proof of the Meteoric Constitution of Saturn's Rings." *Astrophysical Journal* 1 (1895): 416–27.

Kellermann, K., and B. Sheets, eds. *Serendipitous Discoveries in Radio Astronomy: Proceedings of a Workshop Held at the National Radio Astronomy Observatory, Green Bank, West Virginia on May 4, 5, 6, 1983*. Green Bank, W.V.: National Radio Astronomy Observatory, 1983.

Kennicutt, Robert C. "Evolution of Stars and Gas in Galaxies." In *The Astrophysical Journal: American Astronomical Society Centennial Issue*, ed. Helmut A. Abt. Chicago: University of Chicago Press, 1999.

Kenworthy, M. A., and E. E. Mamajek. "Modeling Giant Extrasolar Ring Systems in Eclipse and the Case of J1407B: Sculpting by Exomoons?" *Astrophysical Journal* 800 (2015): 126–35.

Kepler, Johannes. *Epitome of Copernican Astronomy and Harmonies of the World*. Trans. Charles Glenn Wallis. Amherst, N.Y.: Prometheus Books, 1995.

Kidwell, Peggy Aldrich. "Three Women of American Astronomy." *American Scientist* 78 (1990): 244–51.

Kirchhoff, G. *Researches on the Solar Spectrum and the Spectra of the Chemical Elements*. Cambridge: Macmillan and Company, 1862.

Klebesadel, Ray W., Ian B. Strong, and Roy A. Olson. "Observations of Gamma-Ray Bursts of Cosmic Origin Strong." *Astrophysical Journal Letters* 182 (1973): L85–L88.

Kragh, H. "The Beginning of the World: Georges Lemaître and the Expanding Universe." *Centaurus* 32 (1987): 114–39.

Kragh. H. "Georges Lemaître." *Dictionary of Scientific Biography*, vol. 18, suppl. 2. New York: Scribner's, 1990.

Kragh, H. *Conceptions of Cosmos*. Oxford: Oxford University Press, 2007.

Kuchař, Karel V. "Time and Interpretations of Quantum Gravity." In *Proceedings of the 4th Canadian Conference on General Relativity and Relativistic Astrophysics*, ed. G. Kunstatter, D. Vincent, and J. Williams. Singapore: World Scientific, 1992.

Kuchař, Karel V. "Canonical Quantum Gravity." In *General Relativity and Gravitation 1992*, ed. R. J. Gleiser, C. N. Kozameh, and O. M. Moreschi. Bristol, U.K.: Institute of Physics Publishing, 1993.

Landau, L. "On the Theory of Stars." *Physicalische Zeitschrift der Sowjetunion* 1 (1932): 285–88.

Lane, K. Maria D. "Mapping the Mars Canal Mania." *Imago Mundi* 58 (2006): 198–211.

Laplace, P. S. *The System of the World*, vol. II. Trans. J. Pond. London: W. Flint, 1809.

Larson, Richard B., and Linda L. Stryker. "Beatrice Muriel Hill Tinsley." *Quarterly Journal of the Royal Astronomical Society* 23 (1982): 162–65.

Leary, Warren. "Capsule Carrying Interstellar Samples Lands Safely." *New York Times*, January 16, 2006, p. A12.

Leavitt, H. S. "1777 Variables in the Magellanic Clouds." *Annals of the Astronomical Observatory of Harvard College* 60 (1908): 87–108.

Leavitt, H., and E. C. Pickering. "Periods of 25 Variable Stars in the Small Magellanic Cloud." *Harvard College Observatory Circular* 173 (1912): 1–3.

Lemaître, G. "The Beginning of the World from the Point of View of Quantum Theory." *Nature* 127 (1931a): 706.

Lemaître, G. "Contributions to a British Association Discussion on the Evolution of the Universe." *Nature* 128 (1931b): 704–6.

Lemaître, G. "A Homogeneous Universe of Constant Mass and Increasing Radius Accounting for the Radial Velocity of Extra-Galactic Nebulae." *Monthly Notices of the Royal Astronomical Society* 91 (1931c): 483–89.

Lemaître, Canon Georges. *The Primeval Atom.* New York: Van Nostrand, 1950.

Levy, David. *Comets: Creators and Destroyers.* New York: Simon and Schuster, 1998.

Levy, David. *David Levy's Guide to Observing and Discovering Comets.* Cambridge: Cambridge University Press, 2003.

"Lights All Askew in the Heavens." *New York Times*, November 10, 1919, p. 17.

Littmann, Mark. *Planets Beyond: Discovering the Outer Solar System.* New York: John Wiley, 1988.

Lorentz, H. A., A. Einstein, H. Minkowski, and H. Weyl. *The Principle of Relativity.* Trans. W. Perrett and G. B. Jeffery. London: Methuen and Company, 1923.

Lowell, A. L. *Biography of Percival Lowell.* New York: Macmillan, 1935.

Lowell, Percival. *Mars as the Abode of Life.* New York: Macmillan, 1908.

Marcy, Geoffrey W., and R. Paul Butler. "A Planetary Companion to 70 Virginis." *Astrophysical Journal Letters* 464 (1996): L147–L151.

"Mars." *Wall Street Journal*, December 28, 1907, p. 1.

Mather, John C., and John Boslough. *The Very First Light.* New York: Basic Books, 1996.

Matthews, Thomas A., and Allan R. Sandage. "Optical Identification of 3C 48, 3C 196, and 3C 286 with Stellar Objects." *Astrophysical Journal* 138 (1963): 30–56.

Maxwell, James Clerk. *On the Stability of the Motion of Saturn's Rings.* Cambridge: Macmillan and Company, 1859.

Mayor, M., and D. A. Queloz. "Jupiter-Mass Companion to a Solar-Type Star." *Nature* 378 (1995): 355–59.

McCormmach, Russell. "John Michell and Henry Cavendish: Weighing the Stars." *British Journal for the History of Science* 4 (1968): 126–55.

McCrea, W. "Personal Recollections." In *Modern Cosmology in Retrospect*, ed. B. Bertotti et al. Cambridge: Cambridge University Press, 1990.

"Meeting of the Royal Astronomical Society, Friday, 1935 January 11." *The Observatory* 58 (1935): 33–41.

Michell, John. "An Inquiry into the Probable Parallax, and Magnitude of the Fixed Stars, from the Quantity of Light Which They Afford us, and the Particular Circumstances of Their Situation." *Philosophical Transactions of the Royal Society of London* 57 (1767): 234–64.

Michell, John. "On the Means of discovering the Distance, Magnitude, &c. of the Fixed Stars, in consequence of the Diminution of the Velocity of their Light, in case such a Diminution should be found to take place in any of them, and such other Data should be procured from Observations, as would be farther necessary for that Purpose." *Philosophical Transactions of the Royal Society of London* 74 (1784): 35–57.

Millikan, Robert. "Available Energy." *Science* 68 (1928): 279–84.

"Millikan Rays." *New York Times*, November 12, 1925, p. 24.

Minkowski, H. "Space and Time." In Lorentz et al. (1923), pp. 75–91.

Montgomery, Colin, Wayne Orchiston, and Ian Whittingham. "Michell, Laplace and the Origin of the Black Hole Concept." *Journal of Astronomical History and Heritage* 12 (2009): 90–96.

Morgan, William W., Stewart Sharpless, and Donald Osterbrock. "Some Features of Galactic Structure in the Neighborhood of the Sun." *Astronomical Journal* 57 (1952): 3.

Ne'eman, Yuval, and Yoram Kirsh. *The Particle Hunters*. Cambridge: Cambridge University Press, 1986.

"New Radio Waves Traced to Centre of the Milky Way." *New York Times*, May 5, 1933, p. 1.

Newton, Isaac. *The Principia*. Trans. I. Bernard Cohen and Anne Whitman. Berkeley: University of California Press, 1999.

Oates, Whitney Jennings, ed. *The Stoic and Epicurean Philosophers: The Complete Extant Writings of Epicurus, Epictetus, Lucretius, Marcus Aurelius*. New York: Random House, 1940.

Oppenheimer, J. R., and Robert Serber. "On the Stability of Stellar Neutron Cores." *Physical Review* 54 (1938): 540.

Oppenheimer, J. R., and H. Snyder. "On Continued Gravitational Contraction." *Physical Review* 56 (1939): 455–59.

Oppenheimer, J. R., and G. M. Volkoff. "On Massive Neutron Cores." *Physical Review* 55 (1939): 374–81.

Osterbrock, Donald E. *Walter Baade: A Life in Astrophysics*. Princeton, N.J.: Princeton University Press, 2001.

Osterbrock, Donald. *James E. Keeler: Pioneer American Astrophysicist*. Cambridge: Cambridge University Press, 2002.

Ostriker, J. P., and P. J. E. Peebles. "A Numerical Study of the Stability of Flattened Galaxies." *Astrophysical Journal* 186 (1973): 467–80.

Otis, Russell M., and R. A. Millikan. "Minutes of the Washington Meeting: The Source of the Penetrating Radiation Found in the Earth's Atmosphere." *Physical Review* 23 (1924): 778–79.

Overbye, Dennis. "And Now There Are Eight." *New York Times*, August 25, 2006, p. 20.

Overbye, Dennis. "Universe as an Infant: Fatter Than Expected and Kind of Lumpy." *New York Times*, March 22, 2013, p. A10.

Pannekoek, A. *A History of Astronomy*. New York: Dover, 1989.

Payne, Cecilia H. *Stellar Atmospheres: A Contribution to the Observational Study of High Temperatures in the Reversing Layers of Stars.* Cambridge, Mass.: Harvard Observatory Monographs, 1925.

Payne-Gaposchkin, Cecilia. *The Dyer's Hand.* In Haramundanis (1984).

Penrose, Roger. *The Emperor's New Mind.* Oxford: Oxford University Press, 1989.

Penzias, A. A., and R. W. Wilson. "A Measurement of Excess Antenna Temperature at 4080 Mc/s." *Astrophysical Journal* 142 (1965): 419–21.

Perlmutter, S., et al. "Measurements of Ω and Λ from 42 High-Redshift Supernovae." *Astrophysical Journal* 517 (1999): 565–86.

Philip, A. G. Davis, and D. H. DeVorkin, eds. "In Memory of Henry Norris Russell." *Dudley Observatory Report* 13 (1977): 107.

Piazzi, G. *Risultati delle Osservazioni della Nuova Stella* (Observations of the new star). Palermo: 1801. Translated by Antonio Parachinatti for Nevil Maskelyne of the Royal Greenwich Observatory. RGO Manuscript 4/221 in Cambridge University Library.

Pickering, E. C. *Harvard College Observatory Annual Report* 52 (1898): 1–14.

Pollack, James B. "The Rings of Saturn." *Space Science Reviews* 18 (1975): 3–93.

"A Prize for Lemaître." *Literary Digest*, March 31, 1934, p. 16.

"Radio Waves Heard from Remote Space." *New York Times*, May 13, 1933, p. 19.

Randall, Lisa. *Warped Passages.* New York: Ecco, 2005.

Rao, Joe. "The 9 Most Brilliant Comets Ever Seen." Space.com, October 5, 2012, http://www.space.com/17918–9-most-brilliant-great-comets.html. Accessed May 3, 2017.

"Red Shift of Nebulae a Puzzle, Says Einstein." *New York Times*, February 12, 1931, p. 15.

Reich, W., and R. Wielebinski. "The Development of Radio Astronomy." *Astronomische Nachrichten* 323 (2002): 530–33.

Renn, Jürgen, Tilman Sauer, and John Stachel. "The Origin of Gravitational Lensing: A Postscript to Einstein's 1936 *Science* Paper." *Science* 275 (1997): 184–86.

Riess, Adam G., et al. "Observational Evidence from Supernovae for an Accelerating Universe and a Cosmological Constant." *Astronomical Journal* 116 (1998): 1009–38.

Rosenfeld, Albert. "What Are Quasi-Stellars? Heavens' New Enigma." *Life*, January 24, 1964, pp. 11–12.

Rovelli, Carlo. "What Does Present Days Physics Tell Us about Time and Space?" Annual Lecture at the Center for the Philosophy of Science, University of Pittsburgh, September 17, 1993.

Rubin, Vera C., W. Kent Ford Jr., and Norbert Thonnard. "Rotational Properties of 21 Sc Galaxies." *Astrophysical Journal* 238 (1980): 471–87.

Ryle, M. "The Nature of the Cosmic Radio Sources." *Proceedings of the Royal Society of London. Series A, Mathematical and Physical Sciences* 248 (1958): 289–308.

Sagan, Carl. *Cosmos.* New York: Ballantine Books, 1980.

Sandage, A. *Centennial History of the Carnegie Institution of Washington*, vol. 1: *The Mount Wilson Observatory.* Cambridge: Cambridge University Press, 2004.

Schilpp, Paul A., ed. *Albert Einstein: Philosopher-Scientist.* New York: Harper Torchbooks, 1959.

Schmidt, Maarten. "3C 273: A Star-Like Object with Large Red-Shift." *Nature* 197 (1963): 1040.

Schwinger, Julian. *Einstein's Legacy.* New York: Scientific American Books, 1986.

Secundus, C. Plinius. *Pliny's Natural History in Thirty-Seven Books*, Book II. London: George Barclay, 1847–48.

Shapley, Harlow. "Luminosity Distribution and Average Density of Matter in Twenty-Five Groups of Galaxies." *Proceedings of the National Academy of Sciences* 19 (1933): 591–96.

Sheehan, William. *The Planet Mars.* Tucson: University of Arizona Press, 1996.

Shu, Frank H. *The Physical Universe: An Introduction to Astronomy.* Mill Valley, Calif.: University Science Books, 1982.

Siegel, Lee J. "A Hotspot for Powerful Cosmic Rays." July 8, 2014, https://archive.unews.utah.edu/news_releases/a-hotspot-for-powerful-cosmic-rays.

Slipher, V. M. "The Radial Velocity of the Andromeda Nebula." *Lowell Observatory Bulletin* 58, no. 2 (1913): 56–57.

Slipher, V. M. "Spectrographic Observations of Nebulae." *Popular Astronomy* 23 (1915): 21–24.

Slipher, V. M. "The Spectrum and Velocity of the Nebula N.G.C. 1069 (M77)." *Lowell Observatory Bulletin* 80, no. 3 (1917a): 59–62.

Slipher, V. M. "Nebulae." *Proceedings of the American Philosophical Society* 56 (1917b): 403–9.

Slipher, V. M. "Dreyer Nebula No. 584 Inconceivably Distant." *New York Times*, January 19, 1921, p. 6.

Smith, Bradford A., and Richard J. Terrile. "A Circumstellar Disk Around ß Pictoris." *Science* 226 (1984): 1421–24.

Smith, R. W. *The Expanding Universe*. Cambridge: Cambridge University Press, 1982.

Smith, R. W. "Edwin P. Hubble and the Transformation of Cosmology." *Physics Today* 43 (1990): 52–58.

Smith, R. W. "Red Shifts and Gold Medals." In his *The Explorers of Mars Hill*, 43–65. West Kennebunk, ME: Phoenix Publishing, 1994.

"Some Scientific Centres—The Heidelberg Physical Laboratory." *Nature* 65 (1902): 587–90.

"Spiral Arms of the Galaxy." *Sky & Telescope*, April 1952, pp. 138–39.

Strauss, D. "Percival Lowell, W. H. Pickering and the Founding of the Lowell Observatory." *Annals of Science* 51 (1994): 37–58.

Strauss, D. *Percival Lowell*. Cambridge, Mass.: Harvard University Press, 2001.

Struve, Otto. "New Light on the Structure of the Galaxy Gained in 1952." *Leaflet of the Astronomical Society of the Pacific*, no. 285, January 1953, pp. 275–82.

Sullivan, Walter. "Pulsations from Space." *New York Times*, April 14, 1968, p. E18.

Sutton, Christine. *Spaceship Neutrino*. Cambridge: Cambridge University Press, 1992.

Swidey, Neil. "Before the Beginning." *Boston Globe*, May 4, 2014, pp. 20–29.

Thorne, Kip S. *Black Holes & Time Warps: Einstein's Outrageous Legacy*. New York: W. W. Norton, 1994.

Time, March 11, 1966 (cover).

Tinsley, Beatrice M. "Evolution of the Stars and Gas in Galaxies." *Astrophysical Journal* 151 (1968): 547–65.

Tinsley, Beatrice M. "Chemical Evolution in the Solar Neighborhood IV." *Astrophysical Journal* 250 (1981): 758–68.

Tully, R. Brent, et al. "The Laniakea Supercluster of Galaxies." *Nature* 513 (2014): 71–73.

Tyson, J. A. "Dark Matter Tomography." *Physica Scripta* T85 (2000): 259–66.

Van de Hulst, Hendrik C., C. Alex Muller, and Jan H. Oort. "The Spiral Structure of the Outer Part of the Galactic System from the Hydrogen Emission at 21 cm Wavelength." *Bulletin of the Astronomical Institutes of the Netherlands* 12 (1954): 117–49.

van de Kamp, Peter. "Astrometric Study of Barnard's Star from Plates Taken with the 24-Inch Sproul Refractor." *Astronomical Journal* 68 (1963): 515–21.

Van Helden, Albert. "Saturn and His Anses." *Journal for the History of Astronomy* 5 (1974): 105–21.

Waldrop, M. Mitchell. "Inflation and the Mysteries of the Cosmos." *Science* 213 (1981): 121–22.

Walsh, Dennis, Robert F. Carswell, and Ray J. Weymann. "0957+561 A, B: Twin Quasistellar Objects or Gravitational Lens?" *Nature* 279 (1979): 381–84.

Walter, Michael. "From the Discovery of Radioactivity to the First Accelerator Experiments." In *From Ultra Rays to Astroparticles*, ed. B. Falkenburg and W. Rhode. Dordrecht: Springer Science, 2012.

Weber, J. "Evidence for Discovery of Gravitational Radiation." *Physical Review Letters* 22 (1969): 1320–24.

Weinberg, Steven. *The First Three Minutes*. New York: Basic Books, 1977.

Weinreb, S., A. H. Barrett, M. L. Meeks, and J. C. Henry. "Radio Observations of OH in the Interstellar Medium." *Nature* 200 (1963): 829–31.

Weiss, Rainer. "Gravitation Research." Massachusetts Institute of Technology, Research Laboratory of Electronics. *Quarterly Progress Report* 105 (1972): 54–76.

Welther, Barbara L. "The Discovery of Sirius B: A Case of Strategy or Serendipity?" *Journal of the American Association of Variable Star Observers* 16 (1987): 34.

Wheeler, John Archibald. "The Superdense Star and the Critical Nucleon Number." In *Gravitation and Relativity*, ed. Hong-Yee

Chiu and William F. Hoffmann, 195–230. New York: W. A. Benjamin, 1964.

Wheeler, John Archibald. "Our Universe: The Known and the Unknown." *American Scientist* 56 (1968): 1–20.

Wheeler, John Archibald. *A Journey into Gravity and Spacetime.* New York: Scientific American Library, 1990.

Wheeler, John Archibald, and Kenneth Ford. *Geons, Black Holes, and Quantum Foam.* New York: W. W. Norton, 1998.

Williams, R. M. E., et al. "Martian Fluvial Conglomerates at Gale Crater." *Science* 340 (2013): 1068–72.

Wolszczan, A., and D. A. Frail. "A Planetary System Around the Millisecond Pulsar PSR1257+12." *Nature* (1992): 145–47.

Young, Peter, et al. "The Double Quasar Q0957+561 A,B: A Gravitational Lens Image Formed by a Galaxy at z = 0.39." *Astrophysical Journal* 241 (1980): 507–20.

Zwicky, F. "Die Rotverschiebung von extragalaktischen Nebeln [The redshift of extragalactic nebulae]." *Helvetica Physica Acta* 6 (1933): 110–27.

Zwicky, F. "Nebulae as Gravitational Lenses." *Physical Review* 51 (1937a): 290.

Zwicky, F. "On the Masses of Nebulae and of Clusters of Nebulae." *Astrophysical Journal* 86 (1937b): 217–46.

Acknowledgments

First and foremost, I extend my gratitude to *Natural History* for providing me the perfect forum to explore my interests in both astronomy and the history of science. For this opportunity I thank the magazine's chief executive officer, Charles Harris, along with editors Vittorio Maestro and Erin Espelie. As several chapters in this book originated in other periodicals, I am also grateful to additional editors for their thoughtful input at the time the articles were first published: Peter G. Brown at *The Sciences*, Steven J. Marcus and Alice Dragoon at *Technology Review*, Tim Appenzeller at *National Geographic*, Laura Helmuth at *Smithsonian*, John S. Rosenberg at *Harvard* magazine, Alan MacRobert and Robert Naeye at *Sky and Telescope*, and John Pomfret at the *Washington Post*.

My literary agent Peter Tallack at The Science Factory was eminently helpful in shepherding my manuscript to the Yale University Press. There I had the immense pleasure of working once again with Joseph Calamia, who always makes the editing process a delight. More than that, he provided the perfect title for this collection. I must also thank manuscript editor Joyce Ippolito, production editor Mary Pasti, and editorial assistant Eva Skewes for making sure that both the text and the images were polished to perfection.

Kisses to my husband Steve Lowe for his loving support. And, lastly, a big hug to my dog Hubble, who remained loyally by my side as I typed away on my computer.

Index